Language and Godels Theorem:
A Revised Edition

Information Science and Technology

Series Editor

Prof. KC Chen,
National Taiwan University, Taipei, Taiwan

Information science and technology ushers 21st century into an Internet and multimedia era. Multimedia means the theory and application of filtering, coding, estimating, analyzing, detecting and recognizing, synthesizing, classifying, recording, and reproducing signals by digital and/or analog devices or techniques, while the scope of "signal" includes audio, video, speech, image, musical, multimedia, data/content, geophysical, sonar/radar, bio/medical, sensation, etc. Networking suggests transportation of such multimedia contents among nodes in communication and/or computer networks, to facilitate the ultimate Internet. Theory, technologies, protocols and standards, applications/services, practice and implementation of wired/wireless networking are all within the scope of this series. Based on network and communication science, we further extend the scope for 21st century life through the knowledge in robotics, machine learning, cognitive science, pattern recognition, quantum/biological/molecular computation and information processing, biology, ecology, social science and economics, user behaviors and interface, and applications to health and society advance.

- Communication/Computer Networking Technologies and Applications
- Queuing Theory, Optimization, Operation Research, Stochastic Processes, Information Theory, Statistics, and Applications
- Multimedia/Speech/Video Processing, Theory and Applications of Signal Processing
- Computation and Information Processing, Machine Intelligence, Cognitive Science, Decision, and Brian Science
- Network Science and Applications to Biology, Ecology, Social and Economic Science, and e-Commerce

For a list of other books in this series please visit www.riverpublishers.com

Language and Godels Theorem: A Revised Edition

By

Bradley S. Tice
Advanced Human Design, USA

River Publishers

Aalborg

ISBN: 9788792329110

Published, sold and distributed by:
River Publishers
PO Box 1657
Algade 42
9000 Aalborg
Denmark

Tel.: +4536953197
www.riverpublishers.com

Contents

1

Introduction

Kurt Gödel's Incompleteness Theorem changed the way logicians and mathematicians thought about the potential for formalizing all of mathematics, especially David Hilbert's plan to formalize, axioms, all of mathematics. The popular notion of the result of Gödel's incompleteness theorem was to have provided the "awkward possibility that the laws of arithmetic might not supply a definitive answer at all" regarding an axomatic system that accounted for all the arithmatic possibilities (Elwes, 2010: 2).

The principle behind Gödel's Incompleteness Theorem is in essence, a word game in that the logical structure of Gödel's argument is parochial paradoxical sentences, in English, that by their black and white semantic structure give only simple yes or no type answers to questions much deeper. This is the starting point for the question I raised in my mathematics dissertation published in 2008 that takes a language oriented approach to the question of whether Gödel's construction of the aegument is the best or the only type of examination of the possibility of the formalization of mathematics (Tice, 2008). My approach to the question was the focus on the 'semantics' of language, in this case the English language, by the examination of the 'meaning', semantics, of the language used to define the word games used by Gödel.

This monograph is a revised edition of my mathematics dissertation (2008) that includes an unpublished manuscript title *The Nature of the Problem*, that addresses the 'why' behind the 'what' that was addressed in the dissertation about my 'deconstruction' of Gödel's approach to Hilbert's axiomatic system of formalization of mathematics. A copy of my mathematics dissertation *Language and Gödel' Theorem* can be found in Appendix A.

Bradley S. Tice, Language and Godels Theorem: A Revised Edition, 1–16.
© 2013 *River Publishers. All rights reserved.*

1.1 The Incompleteness Theorem

The 'Incompleteness Theorem' is the theorem proved by Gödel that no formalized axiomatic system of Peano arithmetic can ever be complete in the sense that some propositions in number theory may never be able to be proved or disproved (Benacerraf and Putnam, 1983). It is the language used, specifically the semantic portion of language use, by Gödel to 'prove' that a consistent systematic axiomatic model for Peanut arithmetic for a foundation for mathematic and logic was untenable.

It was Gödel's Incompleteness Theorem that haunted David Hilbert's attempt to 'axiomatize' all of mathematics and logic into a formalized and foundational approach to mathematics and logic.

Eves and Carroll (1958) states the David Hilbert started The Formalist School after completing his "Grundlagen der Geometrie" (1899) that used axioms from Euclid to develop his formal system for mathematics and logic (Eves and Carroll, 1958: 290). Although Hilbert mentioned his intent to 'develop' The Formalist School in 1904, it was not until the 1920's that Hilbert and other mathematicians would actively develop the work necessary for the discipline to grow (Eves and Carroll, 1958: 290). Hilbert felt that theoretical logic was an 'extension' of The Formal method of mathematics to the 'object' of logic (Carruccio, 1964: 340).

Hilbert's ideas on mathematical formalism stem from Peano's philosophy of mathematical logic that "studies the properties of the operations and the relations of logic" (Carruccio, 1964: 340). Peano continues "the object is to formulate the simplist system of logic notions that are necessary and sufficient symbolically for mathematical truths and their proofs" (Carruccio, 1964: 340). Hilbert considered his formal system as a "Proof Theory" that would allow a 'consistency' to mathematics and logic (Eves and Carroll, 1958: 290).

1.2 Hilbert's Axiomatic System for Mathematics

Before starting with Hilbert's axiomatic system for mathematics some attention will be directed to what lead up to Hilbert and his proposal for such a 'quest' at the Second International Congress of Mathematics in Paris in 1900 (Reid, 1970: 74). During the late 19th Century mathematicians and philosophers like Georg Canter and Bertrand Russell tried to put mathematics on a formalized foundation from which to build (Casti, 2001: 77). But it was David Hilbert that lead the charge to formalization into the 20th Century with the Paris conference in 1900.

In Constance Reid's book *Hilbert* (1970) she notes that Hilbert's address at the 1900 conference was very 'organic' and 'nature' oriented with allusions to trees and ecological systems (Reid, 1970: 77). Hilbert states: "that it shall be possible to establish the correctness of the solution by means of a finite number of steps based upon a finite number of hypothesizes which are implied in the statement of the problem" (Reid, 1970: 77). These axioms as Drucker notes "can be considered definitions for the terms used in them and that the form of the axioms mattered, not the related objects defined by the axioms (Drucker, 1995: 926). Philosophically, Hilbert was at odds with Frege who stated that the "theorems are true of the objects defined by the axiom system" but only in 'proving' rather than a simplistic, and potentially misleading, answer of 'true' (Gray, 2000: 102). Hilbert's mathematics allowed for contradiction in sets of objects, but separately, not simultaneously (Gray, 2000: 102–103 and Note #1)).

The area of philosophy of interest to this monograph is the development of 'paradoxes and word problems' and hence, the world of mathematics and logic of the late 19th and early part of the 20th Century. Garciadiego (1992) notes that the new 'semantic paradoxes' of such scholar's as Berry, Zermelo and others were developed with Russell, rather than Russell being just a 'guide' (Garciadiego, 1992: 131). It was Berry's non-logical paradox of 1904 that was communicated to Russell that the paradox of a set of all sets are not members of that set (Garciadiego, 1992: 136).

It was Gödel's 1931 paper that used these 'paradox puzzles' to show holes in the continuum hypothesis that underlies the universal and complete systemization of the foundations of mathematics and logic. Gray (2000) notes Gödel's interest in the semantic aspects of philosophical truths and that Gödel's completeness theorem was a "semantically complete" model (Gray, 2000: 168–169). Wang (1987) comments "that in 1930 Gödel had known that truth n language cannot be defined itself and that truth differs from probability" (Wang, 1987: 85). It is also these scmatic paradoxes that undo Gödel's argument if they are modified, de-constructed, by the changing of one or two words in each of the paradoxes sentences. This is the topic of my mathematics dissertation (2008) on *Language and Gödel's Theorem* and the reason for this revised account in this monograph.

1.3 Of Two Words

In my mathematics dissertation 'Language and Gödel's Theorem' the change of the word 'will' with the word 'may' to a modification of Gödel's incompleteness theorem, by Rucker (1982) that gives the following:

A Universal Truth Machine, UTM, will never say G is true (Tice 2008: 6).

A Universal Truth Machine, UTM, will say G is true (Tice, 2008: 6).

My thesis is that by replacing the words 'will never' in the first sentence and 'will' in the second sentence with the following:

A Universal Truth Machine, UTM, may never say G is true (Tice, 2008: 6).

A Universal Truth Machine, UTM, may say G is true (Tice, 2008: 6).

I conclude that this replacement of words changes the fundamental nature of the outcome of Gödel's word problem.

I began the structured 'de-construction' of Gödel's 'Incompleteness Theorem' by addressing the simple 'true' and 'false' nature of the semantic paradoxes used by Gödel. It was important to remove the simplistic 'black' and 'white' binary nature of the sentences to a more qualitative answer than a rigid 'yes' or 'no' type of response. I used Rucker's (1982) version of Gödel's semantic paradox sentences in the form of a 'Universal Truth Machine', or UTM, that would provide an ideal set of paradox sentences from which to replace 'key' morphemes, words, to change the semantic nature of the sentence's questions. The basic notion behind the use of English as a natural language is that the transfer of concepts found in mathematical formulas must equal the semantic language found in the word representation of the mathematics used (Tice, 2008: 9).

In reviewing the new sentences from *Language and Gödel's Theorem* (2008) both

"A Universal Truth Machine, UTM, may never say G is true."

And

"A Universal Truth Machine, UTM, may say G is true."

The addition of the word 'may' changes the very nature of the semantic value of each sentence as the word 'may' implies a broader 'inclusion' of concepts than does the original 'absolute' boundaries set by the word 'will'(Tice, 2008: 6). The relatively minimal change of one word added and one word removed has had a significant change to the very meaning of the whole sentence and provides a more accurate tool for the evaluation of a metric to judge valuations found at the foundational levels for both mathematic and logic.

1.4 Language and Gödel's Theorem

The language, or semantic model, of examination of Gödel's incompleteness theorem is the primary area of interest for this monograph and raises the question of the true power of the semantic model of analysis. In reviewing the systematic approach used in my mathematics dissertation (2008) to build a structured 'de-construction' of Gödel's theorem based on a 'semantic' model of interpretation of the word game used by Gödel (Tice, 2008).

I examine the Liar's Paradox in my mathematics dissertation (2008) and use the following two sentences as samples of transfers of 'key' words, the adding of the word 'may' to each original sentence, and the removing of the original word 'is' in each original sentence, to change the nature and meaning of each original sentence (Tice, 2008: 10–11):

Original Sentences:

1. This sentence is not true.
2. This sentence is not provable.

Each of these original sentences are inconsistent with both the Law of Contradiction and the Law of the Excluded Middle (Tice, 2008: 12).

The Law of Contradiction states that a sentence must not be both itself and a contradiction of itself (Tice, 2008: 12).

The Law of the Excluded Middle states that either the sentence holds or a denial of the sentence holds, but not both (Tice, 2008: 12).

Modified Sentences:

1a.) This sentence may be true.
2a.) This sentence may be provable.

These modified sentences allows for more of a degree of inclusion without being contradictory, in that either a true or a false answer can result but not both, and it also obeys the excluded middle law in that either the sentence is true or the sentence is false but not both qualities at the same time because of the semantic value of the word 'may' that allows for such interpretations (Tice, 2008: 12).

Because both the Law of contradiction and The Law of the Excluded Middle are satisfied the melding qualities of the word 'may' in the revised sentences of the Liar's Paradox, the 'de-construction' of Gödel's Theorem results because the use of the word 'may' that is inclusive verses the word 'will' and 'will not' that are regimented and narrow as a semantic quality with corresponding 'either' and 'or' binary type answers the result.

The resulting effect on Gödel's 'Incompleteness Theorem' is that it is now possible to have a consistency within logic and mathematics, without the need for a level of absolutes that are a rarity in systematic valuations, and retains a high level of functionality towards 'understanding' mathematics and logic, rather than 'wishful' grammars, or laws, that are artificial at best, and do little to develop mathematics and logic beyond the dogma of philosophy.

1.5 Can Machines Think?

This chapter is a bit of a divergence from the main point of this topic, but has a common union with the development and advent of the computer by way of the history of Gödel's paper and early computer scientists, actually mathematicians, that found the systematic logic found in Gödel's work to have foundational attributes to the design of the earliest, modern, 20th Century, computer theories. The most promenent of them the Church-Turing theory on computing.

In 2011 IBM's Watson super computer was placed against the television game show 'Jeopardy' that quizes the contestants, and the computer, on their knowledge of 'facts' (Stanley, 2011: 1–2, Markoff, 2011: 1–5 and Ante, 2011: 1–2). The computer won which is not surprising in that it has few limits to reliable resources, no need to sleep, no emotions, regular physical states, in other words an electronic encyclopedia. Many philosophers have asked such questions in different ways but it was Turing's 1950 paper 'Computing Machinery and Intelligence' that has been the model for such questioning to this day (Tice, 2005a). Some early paper's have clearly stated 'no' to such a question in an artificial machine, like the scientist interviewed for Feigenbaum and Feldman's 1963 book on computers and thought (Feigenbaum and Feldman, 1963: 2–3). Perhaps the most well known paper to question machine intelligence is the philosopher J.R. Lucas (1961) who notes that human 'reasoning' can recognize the truth of a sentence, but a logical system, a machine, can not (Crevier, 1993: 22).

While the field of artificial intelligence was never an interest to me, I wrote two technical papers for my company, Advanced Human Design, in 2005, on 'A Theory on Neurological Systems- Part I & II that addresses artificial intelligence and mechanical thought (Tice, 2005a and 2005b and Appendix C & D). In these technical reports, I exam the question of machine intelligence from a paper I wrote for a conference in 1997 titled "A Turing Machine: A Question of Linguistics?" that addressed the use of human language as the sole criteria for evaluating a 'thinking like' behavior and that a human that had auditory, visual and voice deficiencies, would be similarly impaired and may fail the test (Tice, 2004: 207–214).

The most interest part of the technical papers is in the first paper, in section Part III, (2005a) that I suggest that numbers and algorithms be treated like von Neumann's concept of 'unreliable' components as found in an inorganic or organic system; organism or machine, that when they become 'problematic' are removed and replaced (Tice, 2005a: 2). This would allow for a more robust and functional system.

Conclusion

The monograph has addressed the processes used in my mathematics dissertation (2008) to de-construct Gödel's semantic paradoxes that he used to prove that a systematic axiomization of Peano arithmetic would be incomplete and inconsistent.

The use of semantic paradoxes allowed the use of the transfer of key words to change the meaning found in the original 1931 paper of Gödel on the incompleteness theorem. By changing the semantics of the paradoxes, the very nature of the questions leads to new interpretations of the answers. The resulting change in meaning changes the nature of Gödel's proof and gives a more viable answer that 'includes' variables in mathematics and logic rather than wishes them away.

Summary

The monograph has filled a void in number theory left by Gödel's' Incompleteness Theorem by allowing for a new interpretation of Gödel's 1931 paper and the true nature of a systematic axiomization of Peano arithmetic to mathematics and logic.

Notes

Note #1 A fundamental axiom in set theory is Zermelo's Axiom of Choice that states "given any set of mutually exclusive nonempty sets, there exists at least one set that contains exactly one element in common with each of the nonempty sets" (Weisstein, 2003: 147). This axiom was related to Hilbert's first problem and in 1963 Cohen demonstrated that this axiom is also independent of the Zermelo-Frankel set theory (Weisstein, 2003: 147).

References

[1] Ante, S.E. (2011) "Computer thumps 'Jeopardy' minds". The Wall Street Journal, February 17, 2011. Pp. 1–2. Website: http://online.wsj.com/article/SB10001424052748704171004576148974172060658.html?

[2] Benacerraf, P. and Putnam, H. (1983) *Philosophy and Mathematics*. Cambridge: Cambridge University Press.

[3] Casti, J.L. (2001) *Mathematical Mountaintops*. Cambridge: Oxford University Press.

[4] Crevier, D. (1993) *AI*. New York: Basic Books.

[5] Carruccio, E. (1964) *Mathematics and Logic in History and in Contemporary Thought*. London: Faber and Faber.

[6] Eves, H. and Newson, C.V. (1958) *An Introduction to the Foundations and Fundamental Concepts of Mathematics*. New York: Holt, Rinehart and Winston.

[7] Drucker, T. (1995) "David Hilbert". In *Notable Twentieth-Century Scientists*,. Volume 2 F-K. Edited by E.J. McMurray, pp. 925–927. New York: Gale Research Inc.

[8] Elwes, R. (2010) "To infinity and beyond: The struggle to save arithmetic". New Scientist, Issue 2773, August 16, 2010, pp. 1–6.

[9] Web site: http://www.newscientist.com/article/mg20727731.300-to-infinity-and-beyond-the-struggle

[10] Feigenbaum, E.A. and Feldman, J. (1963) *Computers and Thought*. New York: McGraw-Hill Book Company.

[11] Garciaadiego, A. R. (1992) *Bertrand Russell and the Origins of the Set-theoretic 'Paradoxes'*. Berlin: Birkhauser Verlag.

[12] Gray, J.J. (2000) *The Hilbert Challenge*. Cambridge: Oxford University Press.

[13] Markoff, J. (2011) "A fight to win the future: Computers vs. Humans". February 14, 2011, pp. 1–5. Website: http://www.nytimes.com/2011/02/15/science/15essay.html?r=l&hpw=&pagewanted=print

[14] Reid, C. (1970) *Hilbert*. Berlin: Springer-Verlag.

[15] Rossberg, M. (2010) "Logical Arithmetic". Research Topic from Mathematics Project. May 2, 2010.

[16] Stanley, A. (2011) "The TV watch: Man, Machine and the trivial pursuit". The New York Times, February 15, 2011, pp. 1–2. Website: http://artsbeat.blogs.nytimes.com/2011/-2/15/the-tv-watch-man-machine-and-trivial-pursuit

[17] Tice, B.S. (2004) *Thought, Function and Form: The Language of Physics*. Bloomington, 1st Books Library.

[18] Tice, B.S. (2005) "A theory on neurological systems-Part I". *Technical Report*, Advanced Human Design, Volume 1, Number 1, October 2005, pp. 1–3.

[19] Tice, B.S. (2005) "A theory on neurological systems-Part II". *Technical Report*, Advanced Human Design, Volume 1, Number 2, November 2005, pp. 1–2.

[20] Tice, B.S. (2008) *Language and Gödel's Theorem*. Aachen: Shaker Verlag.

[21] Weisstein, E.W. (2003) *CRC Concise Encyclopedia of Mathematics*. New York: Chapman & Hall/CRC.

[22] Wang, H. (1987) Reflections on Kurt Gödel. Cambridge: The MIT Press.

Appendix

A.1 Introduction

Gödel's use of the liar's paradox is a semantic valuation of a mathematical problem of the uncertainty of Hilbert's axiomatic system. The language used in Gödel's 'Incompleteness Theorem' is the very bases of this logical paradox. To change the very nature of the black and white answers that result from this type of language game, that result in the simple true/false valuations, an example is made were a Universal Truth Machine does become truly universal.

A.2 Language and Gödel's Theorem

Gödel's use of the liar's paradox is a semantic valuation of a mathematical problem of the uncertainty of Hilbert's axiomatic system. The language used in Gödel's 'Incompleteness Theorem' is the very bases of this logical paradox. To change the very nature of the black and white answers that result from this type of language game, that result in the simple true/false valuations, an example is made were a Universal Truth Machine does become truly universal.

Hilary Putnam makes it clear that 'Gödel's Theorem' did not propose an 'absolutely undecidable' factor to Hilbert's axiomatic foundations [from Putnam's paper "Mathematics without Foundations"] [1].

> Strictly speaking, all Gödel's theorem shows is that, in any particular consistent axiomatizable extension of certain finitely axiomatizable subtheories of Peano arithmetic, there are propositions of number theory that can neither be proved nor disproved. It does not follow that any proposition of number theory is, in some sense, absolutely undecidable. However, it may be the case that some proposition of elementary number theory is neither provable nor refutable in any system whose axioms rational debings will ever

have any good reason to accept. This has caused some to doubt whether every mathematical proposition, or even every proposition of the elementary theory of numbers, can be thought of as having a truth value [2].

Rucker, in his *Infinity and the Mind* (1982) gives a basic procedure for Gödel's Incompleteness Theorem [3]:

1. Someone introduces Gödel to UTM, a machine that is suppose to be a Universal Truth Machine, capable of correctly answering any question at all.
2. Gödel asks for the program and circuit diagrams of the UTM. The program may be complicated, but it can only be finitely long. Call the program P(UTM) for Program of the Universal Truth Machine.
3. Smiling a little, Gödel writes out the following sentence: "The machine constructed on the basis of the program P(UTM) will never say that this sentence is true." Call this sentence G for Gödel. Note that G is equivalent to "UTM will never say G is true."
4. Now Gödel laughs his high laugh and asks UTM whether G is true or not.
5. If UTM says G is true, then "UTM will never say G is true" is false. If "UTM will never say G is true" is false, the G is false (since G = "UTM will never say G is true.") So if UTM says that G is true, then G is in fact false, and UTM has made a false statement. So UTM says that G is true, then G is in fact false, and UTM has made a false statement. So UTM will never say that G is true, since UTM makes only true statements.
6. We have established that UTM will never say G is true. So "UTM will never say G is true" is in fact a true sentence. So G is true (since G = "UTM will never say G is true.").
7. "I know a truth that UTM can never utter," Gödel says. "I know that G is true. UTM is not truly universal."

An interesting twist to Gödel's 'Incompleteness Theorem' is the very bases of this logical paradox is to change the very nature of the black and white answers that result from this type of language game. Instead of using simple true/false valuations that are the result of the type of statements being made, if we were to exchange the following statements for what the Universal Truth Machine (UTM) is asked to answer:

1. Replace the statement: "UTM will never say G is true." with "UTM may never say G is true."

2. Replace the statement: "UTM will say *G* is true." with "UTM may say *G* is true."

By replacing the adverb 'never' with the auxiliary verb 'may', the questions become benign as each statement will always be true. Hence the Universal Truth Machine is then truly universal! Does this undo the the primary strengths found in Gödel's Incompleteness Theorem? Yes. But before we can examine this in more detail the importance of language and the language of mathematics must be defined.

In Alfred Tarski's paper "The Semantic Conception of Truth" he outlines some terms that have relevance to the ideas discussed in this paper. Tarski considers the predicate term 'true' to refer to psychological phenomena such as judgements or beliefs, and to certain physical objects, namely linguistic expressions and specifically to sentences and propositions [4]. All truth, like sentences, must be to a specific language, and must be in context so as to retain the original properties of semantic values [5].

While Tarski first suggests the familiar formula:

1. The truth of a sentence consists in its agreement with, or correspondence to, reality.

   His second sentence is more accurate for this discussion and will be used as a bases for truth:

   - A sentence is true if it designates an existing state of affairs.
     While Tarski finds both of these sentences open to misinterpretation there is a definition that will constrain such errors to a minimum:
   - The problem of the definition of truth obtains a precise meaning and can be solved in a rigorous way only for those languages whose structure has been exactly specified [6].

In other words, knowing the structure of a language will keep the structure in context to other structures, i.e. languages, when transferring a concept or a meaning.

Chaitin has noted that Ramsey has stated "Paradoxes are semantic paradoxes, not logical paradoxes, and so they cause trouble for linguists, but not for mathematics" (Cited from Chaitin from F.P. Ramsey (1925) *Collected Papers: The Foundations of Mathematics*. Littlefield, Adams and Company, Patteson: New Jersey, 1960: pp. 20–21). This is not the case as Gödel's use of the liar's paradox is a semantic valuation of a mathematical problem of the uncertainty of Hilbert's axiomatic system. Let use use the following equation set up in a formal mathematical form:

$$1 + 1 = 2 \tag{1.1}$$

We can rewrite this equation using the English language as follows:

One plus one equals two.
$$(1 + 1 = 2) \tag{1.2}$$

They represent the same quantities, but are represented by different language systems.

Now if we made a mistake in coding from the formal mathematic language to that of the English language as follows

$$1 + 1 = 2 \tag{1.3}$$

Becomes:

One plus one equals three.
$$(1 + 1 = 3) \tag{1.4}$$

The sentence is correct in a grammatical sense, but the content desired, the net result of the two 1's added together to become a total of 2 has resulted in the mistake of this being changed to a three. The error is coded in the English language, and is correct from a linguistic point, i.e. grammar and syntax, but from a mathematical stand point, such an equation can only exist in a surreal world far beyond the logic of modern mathematics.

In other words, the linguistic problem with the English sentence becomes a mathematical problem as $1 + 1 = 3$ is false on a mathematical level and on a translational level, i.e. the transition from the numerical equation, $1 + 1 = 2$, to the English language sentence resulted in the mistaken deletion of the 2 and the addition of the the English word three in its place. In other words the importance of language to mathematics is that it is a symbolic representation, i.e. a type of language code, that must be an accurate representation of that mathematical model to be represented.

Gödel's theorem is derived from the 'liars paradox' which states:

(S1) This sentence is not true.

If S1 is true, then what it asserts must be the case, so it is false. Yet if S1 is not true, then it asserts what is the case, so S1 is true. Tarski formalized this paradox in the form of arithmetic and, ultimately, resolved the paradox by concluding that sufficiently rich languages cannot formalize the concept of truth for their sentences [7].

Gödel tried to remove the paradox by stating:

(S2) This sentence is not provable.

If S2 were true, then it would be unprovable and the underlying formal system would be incomplete. If S2 yields a false statement of arithmetic then S2 is provable and results in S2 proof that asserts that [not] S2 is provable leading to an inconsistency [8].

In the Berry paradox:

(P1) The least number not denoted by   a phrase with fewer than fourteen words.

Given a fixed and finite vocabulary, there can be only a finite number of phrases with fewer than fourteen words. P1 picks, [n]; not denoted in the number of phrases, but because P1 has only thirteen words, it cannot denote [n]. Again the Berry paradox is resolved by claiming that sufficiently rich languages can formalize the concept of denotation between terms and numbers [9].

If we revise both S1 and S2 in the same way as I proposed for earlier in this paper, i.e change the word 'will' to that of 'may', the following results:

Original S1:

(S1) This sentence is not true.

Original S2:

(S2) This sentence is not provable.

Revised S1:

This sentence may be true.

Revised S2:

This sentence may be provable.

Note that in revising the terminology used in these sentences we have removed the contradictory aspects of the sentence's semantic values and added a degree of 'chance' into the problem without resulting in a contradiction. Both of these revised sentences adhere to the consistency of proofs of axiomatic systems in that they satisfy both The Law of Contradiction and The Law of the Excluded Middle. An axiom is a self-evident truth.

The Law of Contradiction states that if S is any statement, then the Law of Contradiction states that S and a contradiction (i.e. any denial) of S cannot both hold [10].

14

The Law of the Excluded Middle states that either S holds or a denial of S holds [11].

In examining both of the revised sentences the following applies:

Revised S1:
This sentence may be true.

This sentence follows The Law of Contradiction in that neither S nor a contradiction of S results from the sentence in that the word 'may' allows for either answer, but not both, in the semantic value of the sentence.

Again The Law of the Excluded Middle is followed in that either S holds or a denial of S holds and the use of the word 'may' allows for this interpretation.

Revised S2:
This sentence may be provable.

The same applies for the revised S2 in that it follows both The Law of Contradiction and The Law of the Excluded Middle.

Original S2:
(S2) This sentence is not provable.

If S2 were true, then it would be unprovable and the underlying formal system would be incomplete. If S2 yields a false statement of arithmetic then S2 is provable and results in S2 proof that asserts that [not] S2 is provable leading to an inconsistency [12].

In revising the S2 we have the following:

Revised S2:
This sentence may be provable.

If S2 were true, then it would be provable and the underlying formal system would be complete. If S2 yields a false statement of arithmetic then S2 maybe provable but does not result in specifically asserting that [not] S2 is provable, leading to a consistency in that the sentence stated that it 'may be' provable, but not 'will' be provable.

A.3 Summary

It has been shown that the language used in Gödel's 'Incompleteness Theorem' is the very bases of the logical paradox when Gödel uses the liar's paradox as a semantic valuation of the mathematical problem of the uncertainty found in Hilbert's axiomatic system. When the language is changed, the very nature of the original 'black and white' answers changes the simplistic true and false valuations that have a direct result in altering the very nature of Gödel's theorem.

References

[1] P. Benacerraf and H. Putnam."Philosophy of Mathematics". Cambridge: Cambridge University Press. Cambridge. 1983.

[2] P. Benacerraf and H. Putnam."Philosophy of Mathematics". Cambridge: Cambridge University Press. Cambridge. 1983.

[3] R. Rucker "Infinity and the Mind". Birkhauser. Boston. 1982.

[4] S.R. Givant and R.N. McKenzie."Alfred Tarski: Collected Papers". Birkhauser. Boston. 1986.

[5] S.R. Givant and R.N. McKenzie."Alfred Tarski: Collected Papers". Birkhauser. Boston. 1986.

[6] S.R. Givant and R.N. McKenzie."Alfred Tarski: Collected Papers". Birkhauser. Boston. 1986.

[7] G.J. Chaitin. Information-theoretic computational complexity and Gödel's theorem and information in T. Tymoczko's "New Directions in the Philosophy of Mathematics". Birkhauser. Boston. 1974/1986.

[8] G.J. Chaitin. Information-theoretic computational complexity and Gödel's theorem and information in T. Tymoczko's "New Directions in the Philosophy of Mathematics". Birkhauser. Boston. 1974/1986.

[9] G.J. Chaitin. Information-theoretic computational complexity and Gödel's theorem and information in T. Tymoczko's "New Directions in the Philosophy of Mathematics". Birkhauser. Boston. 1974/1986.

[10] J.R. Newman."The World of Mathematics". Simon and Schuster. New York. 1956.

[11] J.R. Newman."The World of Mathematics". Simon and Schuster, New York. 1956.

[12] G.J. Chaitin. Information-theoretic computational complexity and Gödel's theorem and information in T. Tymoczko's "New Directions in the Philosophy of Mathematics". Birkhauser. Boston. 1974/1986.

[13] K. Gödel "The Completeness of the Axioms of the Functional Calculus of Logic". In Jean van Heijenoort's (1967). *From Frege to Gödel*. Cambridge: Harvard University Press, pages 582–617.

[14] K. Gödel "Some Metamathematical Results on Completeness and Consistency, On. Formally Undecidable Propositions of Principia Mathematical and related Systems-I, and on Completeness and Consistency". In Jean van Heijenoort's (1970) *Frege and Gödel*. Cambridge: Harvard University Press, pages 83–108.

2

The completeness of the axioms of the functional calculus of logic[1]

Kurt Gödel
(1930a)

In his doctoral dissertation at the University of Vienna (*1930*)[a] Gödel proved that the predicate calculus of first order is complete, in the sense that every valid formula is provable. The present paper is a rewritten version of this dissertation.

At the beginning of the paper Gödel mentions Whitehead and Russell's method of deriving logic and mathematics from axioms by means of purely formal rules, and he writes: "Of course, when such a procedure is followed the question at once arises whether the initially postulated system of axioms and principles of inference is complete". One must acknowledge that White-head and Russell had not shown much concern for that problem, any more than they did in general for questions that, being semantic in character, went beyond provability in their system.

The statement that the pure predicate calculus of first order is complete, that is, that every valid formula is provable, is equivalent to the statement that every formula is either refutable or satisfiable. Gödel actually proves a stronger statement, namely, that every formula is either refutable or $\aleph_0$-satisfiable. Hence his proof yields, besides completeness, the Löwenheim Skolem theorem, which states that a satisfiable formula is $\aleph_0$-satisfiable. The proof makes use of a number of devices introduced by Löwenheim (*1915*) and Skolem (*1920*; see also *1922*, *1928*, and *1929*) but contains a step (Theorem VI) through which semantic arguments are connected with provability

---

[1] I am indebted to Professor H. Hahn for several valuable suggestions that were of help to me in writing this paper.

[a] The degree was granted on 6 February 1930.

Bradley S. Tice, Language and Godels Theorem: A Revised Edition, 17–30.

in a definite system. At about the same time, Herbrand, too, was expanding Löwenheim's and Skolem's work; the relation between Gödel's methods and results and those of Herbrand (*1930*) was brought out by Dreben (*1952;* see also above, pp. 510 and 579).

Gödel generalizes his result—that every formula is either refutable or satisfiable—in two directions, to the predicate calculus of first order with identity (in which some irrefutable formulas are finitely satisfiable without being $\aleph_0$-satisfiable) and to infinite sets of formulas. The second generalization (Theorem IX) is derived from the result now known as the finiteness, or compactness, theorem (Theorem X), of which Gödel gives a semantic proof.

By using the procedure of arithmetization introduced by Gödel in another paper (*1931*), Hilbert and Bernays (*1939*, pp. 205–253) were able to give what they call a "proof-theoretic" version of Gödel's completeness theorem: if a formula is irrefutable in the pure predicate calculus of first order, it is irrefutable also in every consistent system S that remains consistent when the axioms of number theory, as well as any verifiable formulas of the theory, are added to the axioms of S (p. 252; see also Theorem 36 in *Kleene 1952*, p. 395).

This proof-theoretic form of Gödel's completeness theorem was extended to the case of infinitely many formulas by Wang (*1951; see* also *1950*, p. 449, Theorem 5, and *1962*, pp. 345–352) and modified so as to give a sharp form of the Löwenheim–Skolem theorem. If Con (Σ) is the usual formula expressing the consistency of a first-order system Σ, the result, called Bernays's lemma by Wang, says that, if we add Con (Σ) to number theory as a new axiom, we can prove in the resulting system arithmetic translations of all theorems of Σ. This lemma has been applied in several directions (see, for instance, *Wang 1952* and *1955*).

The Löwenheim–Skolem theorem shows that, if the predicates assigned to predicate letters in accordance with the definition of validity are just number-theoretic predicates, the class of formulas that turn out to be always true coincides with the class of valid formulas. Kleene (*1952*, pp. 394–395), making explicit some results obtained by Hilbert and Bernays in their arithmetization of Gödel's completeness proof, showed that the predicates can be further restricted to the class $\sum_2 \cap \prod_2$ (in the hierarchy of arithmetic predicates). Putnam (*1961* and *1965*) refined Kleene's result; he restricted the predicates to the class $\sum_1^*$, the smallest class that contains the recursively enumerable predicates and is closed under truth functions; earlier Kreisel (*1953*) and Mostowski (*1955*) had shown that the predicates could not be restricted to the class of recursive predicates, and Putnam (*1956*) that they could not be restricted to the class $\sum_1 \cup \prod_1$.

Another proof of the completeness of first-order logic, along lines somewhat different from those of Gölde proof, was given by Henkin (*1949*).

The translation is by Stefan Bauer-Mengelberg, and it is printed here with the kind permission of Professor Gödel and Springer Verlag.

Whitehead and Russell, as is well known, constructed logic and mathematics by initially taking certain evident propositions as axioms and deriving the theorems of logic and mathematics from these by means of some precisely formulated principles of inference in a purely formal way (that is, without making further use of the meaning of the symbols). Of course, when such a procedure is followed the question at once arises whether the initially postulated system of axioms and principles of inference is complete, that is, whether it actually suffices for the derivation of every true logico-mathematical proposition, or whether, perhaps, it is conceivable that there are true propositions (which may even be provable by means of other principles) that cannot be derived in the system under consideration. For the formulas of the propositional calculus the question has been settled affirmatively; that is, it has been shown[2] that every true formula of the propositional calculus does indeed follow from the axioms given in *Principia mathematica*. The same will be done here for a wider realm of formulas, namely, those of the "restricted functional calculus";[3] that is, we shall prove

Theorem I. *Every valid[4] formula of the restricted functional calculus is provable.*

---

[2] See *Bernays 1926*.

[3] In terminology and symbolism this paper follows *Hilbert and Ackermann 1925*. According to that work, the restricted functional calculus contains the logical expressions that are constructed from propositional variables, $X, Y, Z, \ldots$, and functional variables that is, variables for properties and relations) of type 1, $F(x), G(x,y), H(x,y,z) \ldots$, by means of the operations v (or), $^{-}$ (not), (x) (for all), (Ex) (there exists), with the variable in the prefixes (x) or (Ex) ranging over individuals *only*, *not* over functions. A formula of this kind is said to be valid (tautological) if a true proposition results from every substitution of specific propositions and functions for $X, Y, Z \ldots$, and $F(x), G(x,y), \ldots$ respectively (for example, $(x)[F(x) \vee \overline{F(x)}]$).

[4] To be more precise, we should say "valid in every domain of individuals", which, according to well-known theorems, means the same as "valid in the denumerable domain of individuals". For a formula with free individual variables, $A(x,y, \ldots, w)$, "valid" means that $(x)(y), \ldots, (w)A(x, y, \ldots, w)$ is valid and "satisfiable" that $(Ex)(Ey), \ldots, (Ew)A(x, y, \ldots, w)$ is satisfiable, so that the following holds without exception: "A is valid" is equivalent to "A is not satisfiable".

We lay down the following system of axioms[5] as a basis:

Undefined primitive notions: v, $\overline{}$, and (x). (By means of these, $\&$, $\rightarrow$, $\sim$, and (Ex) can be defined in a well-known way.)

Formal axioms:

1. $XvX \rightarrow X$,
2. $X \rightarrow XvY$,
3. $XvY \rightarrow YvX$.
4. $(X \rightarrow Y) \rightarrow [ZvX \rightarrow ZvY]$,
5. $(x)F(x) \rightarrow F(y)$,
6. $(x)[XvF(x)] \rightarrow Xv\,(x)F(x)$.

Rules of inference:[6]

1. The inferential schema: From A and $A \rightarrow B$, B may be inferred;
2. The role of substitution for propositional and functional variables;
3. From $A(x)$, $(x)A(x)$ may be inferred;
4. Individual variables (free or bound) may be replaced by any others, so long as this does not cause overlapping of the scopes of variables denoted by the same sign.

For what follows, it will be expedient to introduce some abbreviated notations.

(P), (Q), (R), and so on mean prefixes constructed in any way whatever, that is, finite sequences of signs of the form $(x)(Ey)$, $(y)(x)(Ez)(u)$, and the like.

Lower-case German letters, ξ, η, u, v and so on, mean n-tuples of individual variables that is, sequences of signs of the form x, y,z, or x_2,x_1,x_2,x_3, and the like, where the same variable may occur several times. The signs (ξ), $(E\xi)$, and so on are to be understood accordingly. Should a variable occur several times in ξ, we must, of course, think of it as written only once in (ξ) or $(E\xi)$.

Furthermore we require a number of lemmas, which are collected here. The proofs are not given, since they are in part well known, in part easy to supply.

---

[5] It coincides (except for the associative principle, which P. Bernays proved to be redundant) with that given in *Whitehead and Russell 1910*,∗1 and ∗10.

[6] Although Whitehead and Russell use these rules throughout their derivations, they do not formulate all of them explicitly.

1. For every n-tuple ξ

$$(a)(\xi)F(\xi) \rightarrow (E\xi)F(\xi),$$
$$(b)(\xi)F(\xi)\&(E\xi)G(\xi) \rightarrow (E\xi)[F(\xi)\&G(\xi)],$$
$$(c)(\xi)\overline{F(\xi)} \sim \overline{(E(\xi)F(\xi))}$$

are provable.

2. If ξ and ξ' differ only in the order of the variables, then

$$(E\xi)F(\xi) \rightarrow (E\xi')F(\xi)$$

is provable.

3. If ξ consists entirely of distinct variables and if $\xi\prime$ has the same number of terms as ξ, then

$$(\xi)F(\xi) \rightarrow (\xi\prime)F(\xi\prime)$$

is provable, even when a number of identical variables occur in $\xi\prime$.

4. If (p_i) means one of the prefixes (x_i) or (Ex_i) and if (q_i) means one of the prefixes (y_i) *or* (Ey_i), then

$$\left(\sqrt{P_1}\right)\left(\sqrt{P_2}\right)\ldots\left(\sqrt{P_n}\right)F(x_1, x_2, \ldots x_n)$$
$$\&\ (q_1)(q_2)\ldots(q_m)G(y_1, y_2, \ldots y_m)$$
$$\sim (p) \equiv [F(x_1, x_2, \ldots x_n)] \& G(y_1, y_2, \ldots, y_m)$$

is provable[7] for every prefix (P) that is formed from the (p_t) and the (q_i) and satisfies the condition that, for $i < k \leqq n$, (p_i) precedes (p_k) and, for $i < k \leqq m$, (q_i) precedes (q_k).

5. Every expression can be brought into normal form; that is, for every expression A there is a normal formula N such that $A \sim N$ is provable.[8]

6. If $A \sim B$ is provable, so is $\Im(A) \sim \Im(B)$, where $\Im(A)$ represents an arbitrary expression containing A as a part (see *Hilbert and Ackermann 1928*, chap. 3, § 7).

7. Every valid formula of the propositional calculus is provable; that is, Axioms 1–4 form a complete axiom system for the propositional calculus.[9]

---

[7] An analogous theorem holds with V instead of &.

[8] See *Hilbert and Ackermann 1928*, chap. 3, § 8.

[9] See *Bernays 1926*.

We now proceed to the proof of Theorem I and first note that the theorem can also be stated in the following form:

Theorem II. *Every formula of the restricted functional calculus is either refutable[10] or satisfiable (and, moreover, satisfiable in the denumerable domain of individuals).*

That I follows from II can be seen as follows: Let A be a valid expression; then $\overline{A}$, is not satisfiable, hence according to II it is refutable: that is, $\overline{\overline{A}}$ is provable and, consequently, so is A. The converse is as apparent.

We now define a class $\mathfrak{R}$ of expressions K by means of the following stipulations:

1. K is a normal formula;
2. K contains no free individual variable;
3. The prefix of K begins with a universal quantifier and ends with an existential quantifier.

Then we have

Theorem III. *If every $\mathfrak{R}$-expression is either refutable or satisfiable,[11] so is every expression.*

Proof. Let A be an expression not belonging to $\mathfrak{R}$. Let it contain the free variables ξ. As is immediately obvious, the refutability of $(E_\xi)A$ follows from that of A, and conversely (by Lemma 1(c), and either Rule of inference 3 or, for the converse, Axiom 5); the same holds, according to the stipulation in footnote 4, for satisfiability. Let $(P)N$ be the normal form of $(E_\xi)A$, so that.

$$(E_\xi)A \sim (P)N \tag{2.1}$$

is provable. Further let

$$B = (x)(P)(Ey)[N \ \& \sim F(x)va(^c_b)].^{12}$$

Then

$$(P)N \sim B \tag{2.2}$$

---

[10] "A is refutable" is to mean "$\overline{A}$ is provable".

[11] "Satisfiable" without additional specification here and in what follows always means "satisfiable in the denumerable domain of individuals". The same holds for "valid".

[12] The variables x and y must not occur in (P).

is provable (on the basis of Lemma 4 and the provability of $(x)(Ey)[F(x) \vee F(y)]$). B belongs to $\mathfrak{R}$ and thus according to the assumption is either satisfiable or refutable. But, by (1) and (2), the satisfiability of B entails that of $(E_\xi)A$, hence also that of A; the same holds for refutability. Thus A, too. is either satisfiable or refutable. $\qquad\square$

Because of Theorem III, therefore, it suffices to show that
Every $\mathfrak{R}$-expression is either satisfiable or refutable.
For this purpose we define the degree of a $\mathfrak{R}$-expression[13] to be the number of blocks in its prefix that consist of universal quantifiers and are separated from each other by existential quantifiers, and we first prove

Theorem IV. *If every expression of degree k is either satisfiable or refutable, so is every expression of degree $k + 1$.*

Proof. Let $(P)A$ be a $\mathfrak{R}$-expression of degree $K + 1$. Let $(P) = (\xi)(Ea\binom{v}{b}) (Q)$ and let $(Q) = (u)(E\eta)(R)$, where (Q) is of degree k and (R) of degree $k - I$. Further let F be a functional variable not occurring in A. If we now put[14]

$$B = (\xi')(E\eta')F(\xi', \eta') \ \& \ (\xi)(\eta)[F(\xi, \eta) \to (Q)A]$$

And

$$C = (\xi')(\xi)(\eta)(u)(E\eta')(E\eta)(R)\{F(\xi', \eta') \ \& \ [F(\xi, \eta) \to A]\}^{15}$$

then a double application of Lemma 4 in combination with Lemma 6 yields the provability of

$$B \sim C; \tag{2.3}$$

furthermore,

$$B \to (P)A \tag{2.4}$$

is obviously valid. Now C is of degree k and by assumption is therefore either satisfiable or refutable. If it is satisfiable, so is $(P)A$ (by (3) and (4)). If it is refutable, so is B (by (3)); that is, $\overline{B}$ is then provable. In that case, if we substitute $(Q)A$ for F in $\overline{B}$, it follows that

$$\overline{(\xi)(E\eta')(Q)A \ \& \ (\xi)[(Q)A \to (Q)A]}$$

is provable. $\qquad\square$

---

[13] The term "degree of a prefix" is used in the same sense.
[14] An analogous procedure was used by Skolem (*1920*) in proving Löwenheim's theorem.
[15] The variable-sequences $\xi, \xi', \eta, \eta', u, \eta$ are, of course, assumed to be pairwise disjoint.

But since, of course

$$(\xi)(\eta)[(Q)A \to (Q)A]$$

is provable, so is $\overline{(\xi')(E\eta')(Q)A}$; that is, in that case $(P)A$ is refutable. $(P)A$ is there-fore indeed either refutable or satisfiable.

It now remains only to prove

Theorem V. *Every formula of degree 1 is either satisfiable or refutable.*

A few definitions are required for the proof. Let $(\xi)(E_\eta) A(\xi; \eta)$ (abbreviated as $(P)A$) be any formula of degree 1. Let ξ stand for an τ-tuple and η for an s-tuple of variables. We think of the τ-tuples taken from the sequence $x_0, x_1, x_2 \ldots, x_i \ldots$, as forming a sequence ordered according to increasing sum of the subscripts [[and for equal sums according to some convention]]:

$$\xi_1 = (x_0, x_0, \ldots, x_0), \xi_2 = (x_1, x_0, \ldots, x_0), \xi_3 = (x_0, x_1, x_2, \ldots, x_0),$$

and so forth; we now define a sequence $\{A_n\}$ of formulas derived from $(P)A$ as follows:

$$A_1 = A(\xi_1; x_1, x_2, \ldots, x_s),$$
$$A_2 = A(\xi_2; x_{s+1}, x_{s+2}, \ldots, x_2) \,\&\, A_1,$$
$$A_n = A(\xi_n; x_{(n-1)s+1}, x_{(n-1s+2}, \ldots, x_{ns}) \,\&\, A_{x-1}.$$

Let the s-tuple $x_{(n-1)s+1}, \ldots, x_{ns}$ be denoted by η_n, so that we have

$$A_n = A(\xi_n; \eta_n) \,\&\, A_{n-1}. \tag{2.5}$$

Further we define $(P_n)A_n$ by the stipulation

$$(P_n)A_n = (Ex_0)(Ex_1) \cdots (Ex_{ns})A_n.$$

As we can easily convince ourselves, it is precisely the variables x_0 to x_{xs} that occur in A_n; hence they all are bound by (P_n). Further it is apparent that the variables of the τ-tuple ξ_{n+1} already occur in (P_n) (and therefore certainly differ from those occurring in η_{n+1}). Denote by (P_n') what remains of (P_x) when the variables of the τ-tuple ξ_{n+1} are omitted, so that, except for the order of the variables, $(E\xi_{n+1})(P_n') = (P_n)$.

This notation once assumed, we have

Theorem VI. *For every n*

$$(P)A \to (P_n)A_n$$

is provable.

For the proof we use mathematical induction.

I. $(P)A \rightarrow (P_1)A_1$ is provable, for we have

$$(\xi)(E\eta)A(\xi; \eta) \rightarrow (\xi_1)(E\eta_1)A(\xi_1; \eta_1)$$

(by Lemma 3 and Rule of inference 4) and

$$(\xi_1)(E\eta_1)A(\xi_1; \eta_1) \rightarrow (E\xi_1)(E\eta_2)A(\xi_1; \eta_1)$$

(by Lemma 1(a)).

II. For every n, $(P)A \,\#\, (P_n)A_n \rightarrow (P_{n-1})A_{n+1}$ is provable, for we have

$$(\xi)(E\eta)A(\xi; \eta) \rightarrow (\xi_{n+1})(E\eta_{n+1})A(\xi_{n+1}; \eta_{n+1}) \tag{2.6}$$

(by Lemma 3 and Rule of inference 4) and

$$(P_n)A_n \rightarrow (E\xi_{n+1})(P'_n)A_n \tag{2.7}$$

(by Lemma 2). Furthermore,

$$(\xi_{n+1})(E\eta_{n+1})A(\xi_{n+1}; \eta_{n+1}) \ \& \ (E\xi_{n+1})(P'_n)A_n$$
$$\rightarrow (E\xi_{n+1})[(E\eta_{n+1})A(\xi_{n+1}; \eta_{n+1}) \ \& \ (P'_n)A_n] \tag{2.8}$$

(by Lemma 1(b) with the substitutions $(E\eta_{n+1})A(\xi_{n+1}; \eta_{n+1})$ for F and $(P'_n)A_n$ for G).

If we observe that the antecedent of the implication (8) is the conjunction of the consequents of (6) and (7), it is clear that

$$(P)A \ \& \ (P_n)A_n \rightarrow (E\xi_{n+1})[(E\eta_{n+1})A(\xi_{n+1}; \eta_{n+1}) \ \& \ (P'_n)A_n] \tag{2.9}$$

is provable. Furthermore, from (5) and Lemmas 4, 6, and 2 the provability of

$$(E\xi_{n+1})[(E\eta_{n+1})A(\xi_{n+1}; \eta_{n+1}) \ \& \ (P'_n)A_n] \sim (P_{n+1})A_{n+1}] \tag{2.10}$$

follows. II follows from (9) and (10), and from II, together with I, Theorem VI follows.

Assume that the functional variables $F_1, F_2, \ldots, F_k$ and the propositional variables $X_1, X_2, \ldots, X_k$ occur in A. Then A_n consists of elementary components of the form

$$F_1(x_{p1}, \ldots, x_{q1}), F_2(x_{p2}, \ldots, x_{q2}), \ldots, X_1, X_2, \ldots, X_i$$

compounded solely by means of the operations v and $\overline{}$. With each A_n we associate a formula B_n of the propositional calculus by replacing the elementary components of A_n by propositional variables, making certain that different components (even if they differ only in the notation of the individual variables) are replaced by different propositional variables. Furthermore, we understand by "satisfying system of level n [[Erfüllungssystem n-ter Stufe]] of $(P)A$" a system of functions $f_1^{(n)}, f_2^{(n)}, K, f_k^{(n)}$ defined in the domain of integers $z(0 \leq z \leq ns)$ as well as of truth values $w_1^{(n)}, w_2^{(n)}, \ldots, w_l^{(n)}$ for the propositional variables $X_1 \, X_2, \ldots, X_i$ such that a true proposition results if in A_n the F_i are replaced by the $f_i^{(n)}$, the x_i by the numbers i, and the X_i by the corresponding truth values $w_i^{(n)}$. Satisfying systems of level n obviously exist if and only if B_n is satisfiable.

Each B_n, being a formula of the propositional calculus, is either satisfiable or refutable (Lemma 7). Thus only two cases are conceivable:

1. At least one B_n is refutable. Then, as we can easily convince ourselves (Rules of inference 2 and 3, Lemma 1(c)), the corresponding $(P_n)A_n$ is refutable also, and consequently, because of the provability of $(P)A \rightarrow (P_n)A_n$, so is $(P)A$.

2. No B_n is refutable; hence all are satisfiable. Then there exist satisfying systems of every level. But, since for each level there is only a finite number of satisfying systems (because the associated domains of individuals are finite) and since furthermore every satisfying system of level $n + 1$ contains one of level n as a part[16] (as is clear from the fact that the A_n are formed by successive conjunctions), it follows by familiar arguments[16a] that in this case there exists a sequence of satisfying systems $S_1, S_2, \ldots, S_k, \ldots$ (S_k being of level k) such that each contains the preceding one as a part. We now define in the domain of *all* integers ≥ 0 a system $S = \{\varphi_1, \varphi_2, \ldots, \varphi_k; \alpha_1, \alpha_2, \ldots, \alpha_i\}$ by means of the following stipulations:

---

[16] That a system $\{f_1, f_2, \ldots, f_k; w_1, w_2, \ldots, w_i\}$ is part of another, $\{g_1, g_2, \ldots, g_k; v_1, v_2, \ldots v_i\}$, is to mean that

(a) The domain of individuals of the f_i is part of the domain of individuals of the g_i;
(b) The f_i and the g_i coincide within the narrower domain;
(c) For every $i, v_i = w_i$.

[16a] [Apparently by König's infinity lemma (*1926*, p. 120; see also *1927*), which was becoming known among mathematicians at the time Gödel was writing.]

(a) $\varphi_p(a_1, \ldots, a_i)(1 \leqq p \leqq k)$ holds if and only if for at least one S_m of the sequence above (and then for all subsequent ones also) $f_p^{(m)}(a_1, \ldots, a_i)$ holds;

(b) $\alpha_i = w_i^{(m)}(1 \leqq i \leqq l)$ for at least one S_m (and then also for all those that follow). Then it is evident at once that S makes the formula $(P)A$ true. In this case, therefore, $(P)A$ is satisfiable, which concludes the proof of the completeness of the system of axioms given above. Let us note that the equivalence now proved, "valid = provable", contains, for the decision problem, a reduction of the nondenumerable to the denumerable, since "valid" refers to the nondenumerable totality of functions, while "provable" presupposes only the denumerable totality of formal proofs.

Theorem I, as well as Theorem II, can be generalized in various directions. First, it is easy to bring the notion of identity (between individuals) into consideration by adding to Axioms 1–6 above two more:

$$7. \ [x = x, \quad 8. \ [x = y] \to [F(x) \to F(y)].$$

An analogue of what we had above now holds for the extended realm of formulas too:

Theorem VII. *Every formula of the extended realm is provable if it is valid* (more precisely, if it is valid in every domain of individuals), *and. equivalent to VII,*

Theorem VIII. *Every formula of the extended realm is either refutable or satisfiable* (and. moreover, satisfiable in a finite or denumerable domain of individuals).

For the proof, let A denote an arbitrary formula of the extended realm. We construct a formula B as the product (conjunction) of A, $(x)(x = x)$, and all the formulas that we obtain from Axiom 8 by substituting for F the functional variables occurring in A, that is, more precisely,

$$(x)(y)\{x = y \to [F(x) \to f(y)]\}$$

for all singulary functional variables of A

$$(x)(y)(z)\{x = y \to [F(x, z) \to F(y, z)]\} \ \& \ (x)(y)(z)$$
$$(x = y \to [F(z, x) \to F(z, y,)]\}$$

for all binary functional variables of A (including "=" itself), and corresponding formulas for the n-ary functional variables of A for which $n \geq 3$. Let B' be the formula resulting from B when the sign "=" is replaced by a functional variable G not otherwise occurring in B. Then the sign "=" no longer occurs in the expression B' which, therefore, according to what was proved above, is either refutable or satisfiable. If it is refutable, so is B', since it results from B', through the substitution of "=" for G. But B is the logical product of A and a sub formula that is obviously provable from Axioms 7 and 8. In this case, therefore, A is also refutable. Let us now assume that B' is satisfiable in the denumerable domain Σ of individuals for a certain system S of functions.[17] From the way in which B' is formed it is clear that g (that is, the function of the system S that is to be substituted for G) is a reflexive, symmetric, and transitive relation; hence it generates a partition of the elements of Σ, in such a way, moreover, that a function occurring in the system S continues to hold, or not to hold, as the case may be, when elements of the same class are substituted for one another. If therefore, we identify with one another all elements belonging to the same class (perhaps by taking the classes themselves as elements of a new domain of individuals), then g goes over into the identity relation and we have a satisfying system of B, hence also of A. Consequently. A is indeed either satisfiable[18] or refutable.

We obtain a different generalization of Theorem I by considering denumerably infinite sets of logical expressions. For these, too an analogue of Theorems I and II holds, namely

Theorem IX. *Every denumerably infinite set of formulas of the restricted functional calculus either is satisfiable* (that is, all formulas of the system are simultaneously satisfiable) *or possesses a finite subsystem whose logical 'product is refutable, IX follows immediately from*

Theorem X. *For a denumerably infinite system of formulas to be satisfiable it is necessary and sufficient that every finite subsystem be satisfiable,*

Concerning Theorem X we first note that in proving it we can confine ourselves to systems of normal formulas of degree 1, for, by repeated application of the procedure used in the proofs of Theorems III and IV to the individual formulas, we can specify for every system Σ of formulas a system Σ' of

---

[17] If prepositional variables also occur in A. S_i will, of course, have to contain, besides functions, truth values for these prepositional variables.

[18] And, moreover, in an at most denumerable domain (for it consists of disjoint classes of the denumerable domain Σ of individuals).

normal formulas of degree I such that the satisfiability of any subsystem of Σ is equivalent to that of the corresponding subsystem of Σ'.

Thus let

$$(\xi_1)(E\eta_1)A_1(\xi_1; \eta_1), (\xi_2)(E\eta_2)A_2(\xi_2; \eta_2), \ldots, (\xi_\eta)(\xi\eta_n)An(\xi_n; \eta_n), \ldots$$

be a denumerable system Σ of normal expression of degree 1, and let ξ_i be an r_i tuple, η_i an s_r tuple of variables. Let $\xi_1^i, \xi_2^i, \ldots, \xi_n^i, \ldots$ be the sequence of all r_r-tuples taken from the sequence $x_0, x_1, x_2, \ldots, x_n, \ldots$ and ordered according to increasing sum of the subscripts [[and for equal sums according to some convention]]; furthermore, let η_k^i be an s_i tuple of variables,

$$\eta_1^1, \eta_2^1, \eta_3^2, \eta_2^2, \eta_1^3, \eta_4^1, \ldots$$

becomes identical with the sequence $x_1, x_2, \ldots, x_n \ldots$ if every η_k^i is replaced by the corresponding s_i tuple of the x. Further we define, in a way analogous to what was done above, a sequence $\{B_n\}$ of formulas by means of the stipulations

$$B_1 = A_1(\xi_1^1; \eta_1^1)$$

$B_1 = B_{n-1}$ & $A_1(\xi_n^1; \eta_n^1)$ & $A_2(\xi_{n-1}^2; \eta_{n-1}^2)$ & $\ldots$ & $A_{n-1}(\xi_2^{n-1}; \eta_2^{n-1})$ & A_n $(\xi_1^n; \eta_i^n)$. We can easily see that $(P_n)B_n$ (that is, the formula that results from B_n when all individual variables occurring in it are bound by existential quantifiers) is a consequence of the first n expressions of the system Σ given above. If, therefore, every finite subsystem of Σ is satisfiable, so is every B_n. But, if every B_n is satisfiable, so is the entire system Σ(as follows by the argument used in the proof of Theorem V (see p. 588)), and Theorem X is thus proved. Theorems IX and X can be extended without difficulty, by the procedure used in the proof of Theorem VIII, to systems of formulas containing the sign "=".

We can also give a somewhat different turn to Theorem IX if we confine ourselves to systems of formulas without propositional variables and regard them as systems of axioms whose primitive notions are the functional variables occurring in them. Then Theorem IX clearly asserts that every finite or denumerable axiom system in whose axioms "all" and "there exists" never refer to classes or relations but only to individuals[19] either is inconsistent (that is, a contradiction can be constructed in a finite number of formal steps) or possesses a model [[Realisierung]].

---

[19] Hilbert's axiom system for geometry, without the axiom of continuity, can perhaps serve as an example.

Finally, let us also discuss the question of the independence of Axioms 1–8. So far as Axioms 1–4 (those of the propositional calculus) are concerned, it has already been shown by P. Bernays[20] that none of them follows from the other three. That their independence is not affected even by the addition of Axioms 5–8 can be shown by means of the very same interpretations that Bernays uses, provided that, in order to extend them to formulas containing functional variables and the sign "=", we make the following stipulations:

1. The prefixes and individual variables are omitted;
2. In what remains of each formula the functional variables are to be treated just like propositional variables;
3. Only "distinguished" values may ever be substituted for the sign "=".

To demonstrate the independence of Axiom 5, we associate with each formula another one, which we obtain by replacing components of the form

$$(x)F(x), (y)F(y), \ldots; (x)G(x), (y)G(y), \ldots; \ldots,\,^{21}$$

should such occur, by $X \vee \overline{X}$. Then Axioms 1–1 and 6–8 go over into valid formulas, and the same holds, as we can convince ourselves by mathematical induction, of all formulas derived from these axioms by Rules of inference 1–4; Axiom 5, however, does not possess this property. The independence of Axiom 6 can be shown in exactly the same way, except that here $(x)F(x), (y)F(y) \ldots$, and so on must be replaced by $X \,\&\, \overline{X}$. To prove the independence of Axiom 7 we note that Axioms 1–6 and 8 (and therefore also all formulas derived from them) remain valid if the identity relation is replaced by the empty relation, whereas this is not the case for Axiom 7. Similarly, the formulas derived from Axioms 1–7 remain valid when the identity relation is replaced by the universal relation, whereas this is not the case for Axiom 8 (in a domain of at least two individuals). We can also readily see that none of the Rules of inference 1–4 is redundant, but we shall not look into this more closely here.

---

[20] See *Bernays 1926*.

[21] That is, the singular functional variables $F, G \ldots$ preceded by a universal quantifier whose scope is just the $F, G, \ldots$ in question, along with the associated individual variable.

3

Some metamathematical results on completeness and consistency, On formally undecidable propositions of Principia mathematica and related systems I, and On completeness and consistency

Kurt Gödel
(1930b, 1931, and 1931a)

The main paper below (*1931*), which was to have such an impact on modern logic, was received for publication on 17 November 1930 and published early in 1931. An abstract (*1930b*) had been presented on 23 October 1930 to the Vienna Academy of Sciences by Hans Hahn.

Gödel's results are now accessible in many publications, but his original paper has not lost any of its value as a guide. It is clearly written and does not assume any previous result for its main line of argument. It is, moreover, rich in interesting details. We now give some indications of its contents and structure.

Section 1 is an informal presentation of the main argument and can be read by the nonmathematician, it shows how the argument, by dealing with the proposition that states of itself "I am not provable ", instead of the proposition that states of itself "I am not true", skirts the Liar paradox, without falling into it. Gödel also brings to light the relation that his argument bears to Cantor's diagonal procedure and Richard's paradox (Herbrand, on pages 626–628 below, and Weyl (*1949*, pp. 219–235) particularly stress this aspect of Gödel's argument; see also above, p. 439).

Section 2, the longest, is the proof of Theorem VI. The theorem states that in a formal system satisfying certain precise conditions there is an undecidable proposition, that is, a proposition such that neither the proposition itself nor its negation is provable in the system. Before coming to the core of the argument, Gödel takes a number of preparatory steps:

1. A precise description of the system P with which he is going to work. The variables are distinguished as to their types and they range over the natural numbers (type 1), classes of natural numbers (type 2), classes of classes of natural numbers (type 3), and so forth. The logical axioms are equivalent to the logic of *Principia mathernatica* without the ramified theory of types. The arithmetic axioms are Peano's, properly transcribed. The identification of the individuals with the the natural numbers and adjunction of Peano's axioms (instead of their derivation, as in *Principia*) have the effect that every formula has an interpretation in classical mathematics and, if closed, is either true or false in that interpretation; moreover, proofs are considerably shortened.

2. An assignment of natural numbers to sequences of signs of P and a similar assignment to sequences of sequences of signs of P. The first assignment is such that, given a sequence, the number assigned to it can be effectively calculated, and, given a number, we can effectively decide whether the number is assigned to a sequence and, if it is, actually write down the sequence; similarly for the second assignment. By means of these assignments we can correlate number-theoretic predicates with metamathematical notions used in the description of the system; for example, to the notion "axiom" corresponds the predicate $Ax(x)$, which holds precisely of the numbers x that are assigned to axioms (the "Gödel numbers" of axioms, we would say today).

3. A definition of primitive recursive functions (Gödel calls them recursive functions) and the derivation of a few theorems about them. These functions had already been used in foundational research (for example, by Dedekind (*1888*), Skolem (*1923*), Hilbert (*1925, 1927*), and Ackermann (*1928*)); Gödel gives a precise definition of them, which has become standard.

4. The proof that forty-five number-theoretic predicates, forty of them associated with metamathematical notions, are primitive recursive.

5. The proof that every primitive recursive number-theoretic predicate is numeralwise representable in P. That is, the predicate holds of some given numbers if and only if a definite formula of P is provable whenever its free variables are replaced by the symbols that represent these numbers in P.

6. The definition of ω-consistency.

   Gödel can then undertake to prove Theorem VI. The scope of the theorem is enlarged by the addition of any ω-consistent primitive recursive class κ of formulas to the axioms of P. For each such κ a different system is thus obtained (in the present note, "P_κ", a notation not used by Gödel, will denote the system corresponding to a given κ). After the proof Gödel makes a number of important remarks:

(a) He points out the constructive content of Theorem VI.

(b) He introduces predicates that are *entscheidungsdefinit* (in the translation below these are called *decidable* predicates, at the author's suggestion). If we take into account the few lines added in proof at the end of a later note of Gödel's (*1934a*), these predicates are in fact those that today we call recursive (that is, general recursive) predicates. Gödel somewhat extends the result of Theorem VI by assuming only that κ is decidable, and not that it is primitive recursive.

(c) If κ is assumed to be merely consistent, instead of a ω-consistent, the proof yields the existence of predicate whose universalization is not provable but for which no counterexample can be given; P_κ is ω-incomplete, as we would say today.

(d) The adjunction of the undecidable formula Neg(17 Gen r) to κ yields a consistent but ω-inconsistent system.

(e) Even with the adjunction of the axiom of choice or the continuum hypothesis the system contains undecidable propositions.

The section ends with a review of the properties of P that are actually used in the proof and the remark that all known axiom systems of mathematics, or of any substantial part of it, have these properties.

Section 3 presents two supplementary undecidability results. Gödel establishes (Theorem VII) that a primitive recursive number-theoretic predicate is *arithmetic*, that is, can be expressed as a formula of first-order number theory (this yields a stronger result than the numeralwise representability of such predicates, as it was introduced and used in Section 2). Hence every formula of the form $(x)F(x)$, with $F(x)$ primitive recursive, is equivalent to an arithmetic formula: moreover, this equivalence is provable in P_κ: one can review the informal proof presented by Gödel and check that P_κ is strong enough to express and justify each of its steps. Since the proposition that was proved to be undecidable in Theorem VI is of the form $(x)F(x)$, with $F(x)$ primitive recursive. P_κ contains undecidable *arithmetic* propositions (Theorem VIII). For all its strength, the system P_κ cannot decide every first-order number-theoretic proposition. Theorem X states that, given a formula $(x)F(x)$, with $F(x)$ primitive recursive, one can exhibit a formula of the *pure* first-order predicate calculus, say A, that is satisfiable if and only if

$(x)F(x)$ holds. Moreover, since P_κ contains a set theory, the equivalence

$$(x)F(x) \equiv (A \text{ is satisfiable})$$

is expressible in P_κ and, as one can verify by reviewing Gödel's informal argument, provable in P_κ. (Theorem IX) there are formulas of the pure first-order predicate calculus whose validity is undecidable in P_κ.

In Section 4 an important consequence of Theorem VI is derived. The statement "there exists in P_κ an unprovable formula", which expresses the consistency of P_κ, can be written as a formula of P_κ; but this formula is not provable in P_κ (Theorem XI). The main step in the demonstration of this result consists in reviewing the proof of the first half of Theorem VI and checking that all the statements made in that proof can be expressed and proved in P_κ. It is clear that this is the case, and Gödel does not go through the details of the demonstration. The section ends with various remarks on Theorem XI (its constructive character, its applicability to set theory and ordinary analysis, its effect upon Hilbert's conception of mathematics).

Gödel's paper immediately attracted the interest of logicians and, although it caused some momentary surprise, its results were soon widely accepted. A number of studies were directly inspired by it. By using a somewhat more complicated predicate than "is provable in P", Rosser (*1936*) was able to weaken the assumption of w-consistency in Theorem VI to that of ordinary consistency. Hilbert and Bernays (*1939*, pp. 283–340) carried out in all details the proof of the analogue of Theorem XI for two standard systems of number theory, Z_κ and Z, and this proof can be transferred almost literally to any system containing Z. As Gödel, indicates in a note appended to the present translation of his paper, Turing's work (*1937*) gave to the notion of formal system its full generality. The notes of Gödel's Princeton lectures (*1934*) contain the most important results of the present paper, in a more succinct form; they also make precise the notion of (general) recursive function, already suggested by Herbrand (see below, p. 618). In developing the theory of these functions, Kleene (*1936*) obtained undecidability results of a somewhat different character from those presented here. Gödel's work led to Church's negative solution (*1936*) of the decision problem for the predicate calculus of first order. Tarski (*1953*) developed a general theory of undecidability. The device of the "arithmetization" of metamathematics became an everyday tool of the research worker in foundations. Gödel's results, finally, led to a profound revision of Hilbert's program (on that point see, among other texts, *Bernays 1938, 1954* and *Gödel 1958*).

These indications are far from giving a full account of the deep influence exerted in the field of foundations of mathematics by the results presented in the paper below and the methods used to restrict the means of proof in anyway). Hence a consistency proof for the system S can be carried out only by means of modes of inference that are not formalized in the system S itself, and analogous results hold for other formal systems as well, such as the Zermelo-Fraenkel axiom system of set theory.[1]

III. Theorem I can be sharpened to the effect that, even if we add finitely many axioms to the system S (or infinitely many that result from a finite number of them by "type elevation"), we do *not* obtain a complete system, provided the extended system is ω-consistent. Here a system is said to be ω-consistent if, for no property $F(x)$ of natural numbers,

$$F(1), F(2), \ldots, F(n), \ldots \text{ ad infinitum}$$

as well as

$$(Ex)\overline{F(x)}$$

are provable. (There are extensions of the system S that, while consistent, are not ω-consistent.)

IV. Theorem I still holds for all ω-consistent extensions of the system S that are obtained by the addition of *infinitely many* axioms, provided the added class of axioms is decidable [[entscheidungsdefinit]], that is, provided it is metamathematically decidable [[entscheidbar]] for every formula whether it is an axiom or not (here again we suppose that the logic used in metamathematics is that of *Principia mathematica*).

Theorems I, III, and IV can be extended also to other formal systems, for example, to the Zermelo-Fraenkel axiom system of set theory, provided the systems in question are ω-consistent.

The proofs of these theorems will appear in *Monatshefte für Mathematik und Physik*.

A.1 ON FORMALLY UNDECIDABLE PROPOSITIONS OF PRINCIPIA MATHEMATICA AND RELATED SYSTEMS I[1]

(1931)

The development of mathematics toward greater precision has led, as is well known, to the formalization of large tracts of it, so that one can prove any theorem using nothing but a few mechanical rules. The most comprehensive formal systems that have been set up hitherto are the system of *Principia*

---

[1] This result, in particular, holds also for the axiom system of classical mathematics, as it has been constructed, for example, by von Neumann (*1927*).

[2] See a summary of the results of the present paper in *Gödel 1930b*.

mathematica (*PM*)[3] on the one hand and the Zermelo-Fraenkel axiom system of set theory (further developed by J. von Neumann)[4] on the other. These two systems are so comprehensive that in them all methods of proof today used in mathematics are formalized, that is, reduced to a few axioms and rules of inference. One might therefore conjecture that these axioms and rules of inference are sufficient to decide *any* mathematical question that can at all be formally expressed in these systems. It will be shown below that this is not the case, that on the contrary there are in the two systems mentioned relatively simple problems in the theory of integers[5] that cannot be decided on the basis of the axioms. This situation is not in any way due to the special nature of the systems that have been set up but holds for a wide class of formal systems; among these, in particular, are all systems that result from the two just mentioned through the addition of a finite number of axioms,[6] provided no false propositions of the kind specified in footnote 4 become provable owing to the added axioms.

Before going into details, we shall first sketch the main idea of the proof, of course without any claim to complete precision. The formulas of a formal system (we restrict ourselves here to the system *PM*) in outward appearance are finite sequences of primitive signs (variables, logical constants, and parentheses or punctuation dots), and it is easy, to state with complete precision *which* sequences of primitive signs are meaningful formulas and which are not.[7] Similarly, proofs, from a formal point of view, are nothing but finite sequences of formulas (with certain specifiable properties.) Of course, for

---

[3] *Whitehead and Russell 1925*. Among the axioms of the system *PM* we include also the axiom of infinity (in this version: there are exactly denumerably many individuals), the axiom of reducibility, and the axiom of choice (for all types).

[4] See *Fraenkel 1927* and *von Neumann 1925, 1928*, and *1929*. We note that in order to complete the formalization we must add the axioms and rules of inference of the calculus of logic to the set-theoretic axioms given in the literature cited. The considerations that follow apply also to the formal systems (so far as they are available at present) constructed in recent years by Hilbert and his collaborators. See *Hilbert 1922, 1922a, 1927, Bernays 1923, von Neumann 1927*, and *Ackermann 1924*.

[5] That is, more precisely, there are undecidable propositions in which, besides the logical constants $\overline{}$ (not), $\vee$ (or), (x) (for all), and $=$ (identical with), no other notions occur but $+$ (addition) and $\cdot$ (multiplication), both for natural numbers, and in which the prefixes (x), too, apply to natural numbers only.

[6] In *PM* only axioms that do not result from one another by mere change of type are counted as distinct.

[7] Here and in what follows we always understand by "formula of *PM*" a formula written without abbreviations (that is, without the use of definitions). It is well known that [in *PM*] definitions serve only to abbreviate notations and therefore are dispensable in principle.

metamathematical considerations it does not matter what objects are chosen as primitive signs, and we shall assign natural numbers to this use.[8] Consequently, a formula will be a finite sequence of natural numbers,[9] and a proof array a finite sequence of finite sequences of natural numbers. The metamathematical notions (propositions) thus become notions (propositions) about natural numbers or sequences of them;[10] therefore they can (at least in part) be expressed by the symbols of the system *PM* itself. In particular, it can be shown that the notions "formula", "proof array", and "provable formula" can be defined in the system *PM*; that is, we can, for example, find a formula $F(\upsilon)$ of *PM* with one free variable υ (of the type of a number sequence)[11] such that $F(\upsilon)$, interpreted according to the meaning of the terms of *PM*, says: υ is a provable formula. We now construct an undecidable proposition of the system *PM*, that is, a proposition *A* for which neither *A* nor *not-A* is provable, in the following manner.

A formula of *PM* with exactly one free variable, that variable being of the type of the natural numbers (class of classes), will be called a *class sign*. We assume that the class signs have been arranged in a sequence in some way,[12] we denote the *nth* one by $R(n)$, and we observe that the notion " class sign", as well as the ordering relation *R*, can be defined in the system *PM*. Let α be any class sign; by $[\alpha; n]$ we denote the formula that results from the class sign α when the free variable is replaced by the sign denoting the natural number *n*. The ternary relation $x = [y; z]$, too, is seen to be definable in *PM*. We now define a class *K* of natural numbers in the following way:

$$n \varepsilon K \equiv \overline{Bew}[R(n); n] \tag{3.1}$$

[8] That is, we map the primitive signs one-to-one onto some natural numbers. (See how this is done on page 601.)

[9] That is, a number-theoretic function defined on an initial segment of the natural numbers. (Numbers, of course, cannot be arranged in a spatial order.)

[10] In other words, the procedure described above yields an isomorphic image of the system *PM* in the domain of arithmetic, and all metamathematical arguments can just as well be carried out in this isomorphic image. This is what we do below when we sketch the proof; that is, by "formula", "proposition", "variable", and so on. *We must always understand the corresponding objects of the isomorphic image.*

[11] It would be very easy (although somewhat cumbersome) to actually write down this formula.

[12] For example, by increasing sum of the finite sequence of integers that is the "class sign", and lexicographically for equal sums.

(where *Bew x* means: *x* is a provable formula).[13] Since the notions that occur in the definiens can all be defined in *PM*, so can the notion *K* formed from them; that is. there is a class sign *S* such that the formula [*S*; *n*], interpreted according to the meaning of the terms of *PM*, states that the natural number *n* belongs to *K*.[14] Since *S* is a class sign, it is identical with some *R*(*q*); that is; we have

$$S = R(q)$$

for a certain natural number *q*. We now show that the proposition [*R*(*q*); *q*] is un-decidable in *PM*.[15] For let us suppose that the proposition [*R*(*q*); *q*] were provable: then it would also be true. But in that case, according to the definitions given above, *q* would belong to *K*, that is, by (1), $\overline{Bew}$ [*R*(*q*); *q*] would hold, which contradicts the assumption. If, on the other hand, the negation of [*R*(*q*); *q*] were provable, then $\overline{q\varepsilon K}$,[16] that is, *Bew*[*R*(*q*); *q*], would hold. But then [*R*(*q*); *q*], as well as its negation, would be provable, which again is impossible.

The analogy of this argrument with the Richard antinomy leaps to the eye. It is closely related to the "Liar" too;[17] for the undecidable proposition [*R*(*q*); *q*] states that *q*belongs to *K*, that is, by (1), that [*R*(*q*); *q*] is not provable. We therefore have before us a proposition that says about itself that it is not provable [in *PM*].[18] The method of proof just explained can clearly be applied to any formal system that, first, when interpreted as representing a system of notions and propositions, has at its disposal sufficient means of

---

[13] The bar denotes negation.

[14] Again, there is not the slightest difficulty in actually writing down the formula *S*.

[15] "Note that "[*R*(*q*); *q*]" (or, which means the same, "[*S*; *q*]") is merely a *metamathematical description* of the undecidable proposition. But, as soon as the formula *S* has been obtained, we can, of course, also determine the number *q* and, therewith, actually write down the undecidable proposition itself. [This makes no difficulty in principle. However, in order not to run into formulas of entirely unmanageable lengths and to avoid practical difficulties in the computation of the number *q*, the construction of the undecidable proposition would have to be slightly modified, unless the technique of abbreviation by definition used throughout in *PM* is adopted.]

[16] [[The German text reads $\overline{n\varepsilon K}$, which is a misprint.]]

[17] Any epistemological antinomy could be used for a similar proof of the existence of undecidable propositions.

[18] Contrary to appearances, such a proposition involves no faulty circularity, for initially it [only] asserts that a certain well-defined formula (namely, the one obtained from the *q*th formula in the lexicographic order by a certain substitution) is unprovable. Only subsequently (and so to speak by chance) does it turn out that this formula is precisely the one by which the proposition itself was expressed.

expression to define the notions occurring in the argument above (in particular, the notion, "provable formula") and in which, second, every provable formula is true in the interpretation considered. The purpose of carrying out the above proof with full precision in what follows is, among other things, to replace the second of the assumptions just mentioned by a purely formal and much weaker one.

From the remark that $[R(q); q]$ says about itself that it is not provable it follows at once that $[R(q); q]$ is true, for $[R(q);q]$ *is* indeed unprovable (being undecidable). Thus, the proposition that is undecidable *in the system PM* still was decided by meta-mathematical considerations. The precise analysis of this curious situation leads to surprising results concerning consistency proofs for formal systems, results that will be discussed in more detail in Section 4 (Theorem XI).

We now proceed to carry out with full precision the proof sketched above. First we give a precise description of the formal system P for which we intend to prove the existence of undecidable propositions. P is essentially the system obtained when the logic of *PM* is superposed upon the Peano axioms[19] (with the numbers as individuals and the successor relation as primitive notion).

The primitive signs of the system P are the following:

I. Constants: "$\sim$ " (not), "v" (or), "Π" (for all), "0" (zero), "f" (the successor of), "(" , ")" (parentheses);

II. Variables of type 1 (for individuals, that is, natural numbers including 0): "x_i", "y_i","z_i", ...;
Variables of type 2 (for classes of individuals): "x_2", "y_2", "z_2", ...;
Variables of type 3 (for classes of classes of individuals): "x_3", "y_3", "z_3", ...;
And so on, for every natural number as a type.[20]

Remark: Variables for functions of two or more argument places (relations) need not be included among the primitive signs since we can define relations to be classes of ordered pairs, and ordered pairs to be classes of classes; for example, the ordered pair a, b can be defined to be $((a), (a, b))$, where (x, y) denotes the class whose sole elements are x and y, and (x) the class whose sole element is x.[21]

---

[19] The addition of the Peano axioms, as well as all other modifications introduced in the system *PM*, merely serves to simplify the proof and is dispensable in principle.

[20] It is assumed that we have denumerably many signs at our disposal for each type of variables.

[21] Nonhomogeneous relations, too, can be defined in this manner; for example, a relation between individuals and classes can be defined to be a class of elements of the form $((x_2),$

By a *sign of type* 1 we understand a combination of signs that has [any one of] the forms

$$a, \, fa, \, ffa, \, fffa, \, \ldots, \text{ and so on.}$$

where a is either 0 or a variable of type 1. In the first case, we call such a sign a *numeral*. For $n > 1$ we understand by a *sign of type* n the same thing as by a *variable of type* n. A combination of signs that has the form $a(b)$, where b is a sign of type n and a a sign of type $n + 1$, will be called an *elementary formula*. We define the class *of formulas* to be the smallest class[22] containing all elementary formulas and containing $\sim (a)$, $(a)v(b)$, $x\Pi(a)$ (where x may be any variable)[23] whenever it contains a and b. We call $(a)v(b)$ the *disjunction* of a and b, $\sim (a)$ the *negation* and $x\Pi(a)$ a *generalization* of a. A formula in which no free variable occurs (*free variable* being denned in the well-known manner) is called a *sentential formula* [Satzformel]. A formula with exactly n free individual variables (and no other free variables) will be called an *n-place relation sign*; for $n = 1$ it will also be called a *class sign*.

By Subst $a\binom{b}{c}$ where a stands for a formula, v for a variable, and b for a sign of the same type as v) we understand the formula that results from a if in a we replace v, wherever it is free, by b.[24] We say that a formula a is a *type elevation* of another formula b if a results from b when the type of each variable occurring in b is increased by the same number.

The following formulas (I–V) are called *axioms* (we write them using these abbreviations, defined in the well-known manner: $\supset$, $\equiv$, (Ex), $=$,[25] and observing the usual conventions about omitting parentheses):[26]

I.   (a) $\sim (f x_i = 0)$,

$((x_2), x_2))$ Every proposition about relations that is provable in *PM* is provable also when treated in this manner, as is readily seen.

[22] Concerning this definition (and similar definitions occurring below) see *Lukasieuicz and Tarski 1930*.

[23] Hence $x\Pi(a)$ is a formula even if x does not occur in a or is riot free in a. In this case, of course, $x\Pi(a)$ means the same thing as a.

[24] In case v does not occur in a as a free variable we put Subst $a\binom{v}{b} = a$. Note that "Subst" is a meta-mathematical sign.

[25] $x_i = y_i$ is to be regarded as defined by $x_2\Pi(x_2(x_i) \supset x_2(y_i))$, as in *PM* (I, *13) similarly for higher types).

[26] In order to obtain the axioms from the schemata listed we must therefore

1. Eliminate the abbreviations and

(b) $f x_i = f y_i \supset x_i = y_i$,

(c) $x_2(0), x_i \Pi(x_2(x_i) \supset x_2(f x_i)) \supset x_i \Pi(x_2(x_i))$.

II. All formulas that result from the following schemata by substitution of any formulas whatsoever for p, q, r:

(a) $p v p \supset p$,

(b) $p \supset p v q$,

(c) $p v q \supset q v p$,

(d) $(p \supset q) \supset (r v p \supset r v q)$.

III. Any formula that results from either one of the two schemata

(a) $v \Pi(a) \supset \text{Subst } a\binom{\tau}{c}$,

(b) $v \Pi(b v a) \supset b v \tau \Pi(a)$

when the following substitutions are made for a, v, b, and c (and the operation indicated by "Subst" is performed in 1):

For a any formula, for v any variable, for b any formula in which v does not occur free, and for c any sign of the same type as v, provided c does not contain any variable that is bound in a at a-place where v is free.[27]

IV. Every formula that results from the schema

(a) $(Eu)(v \Pi(u(v) \equiv a))$

when for v we substitute any variable of type n, for u one of type $n + 1$, and for a any formula that does not contain u free. This axiom plays the role of the axiom of reducibility (the comprehension axiom of set theory).

V. Every formula that results from

(a) $x_i \Pi(x_2(x_i) \equiv y_2(x_i)) \supset x_2 = y_2$

by type elevation (as well as this formula itself). This axiom states that a class is completely determined by its elements.

A formula c is called an *immediate consequence* of a and b if it is the formula $(\sim (b)) v(c)$, and it is called an *immediate consequence* of a if it is the

---

2. Add the omitted parentheses

(in II, III and IV after carrying out the substitutions allowed).

Note that all expressions thus obtained are "formulas" in the sense specified above. (See also the exact definitions of the metamathematical notions on pp. 603–606.)

[27] Therefore c is a variable or 0 or a sign of the form $f \ldots f u$, where u is either 0 or a variable of type 1. Concerning the notion "free (bound) at a place in a", see I A 5 in *von Neumann 1927*.

formula $v \Pi(a)$, where v denotes any variable. The class of *provable formulas* is defined to be the smallest class of formulas that contains the axioms and is closed under the relation "immediate consequence".[28]

We now assign natural numbers to the primitive signs of the system P by the following one-to-one correspondence;

$$\text{"0"} \dots 1 \quad \text{"}\sim\text{"} \dots 5 \quad \text{"}\Pi\text{"} \dots 9$$
$$\text{"}f\text{"} \dots 3 \quad \text{"}v\text{"} \dots 7 \quad \text{"("} \dots 11$$
$$\text{")"} \dots 13;$$

to the variables of type n we assign the numbers of the form p^n (where p is a prime number > 13). Thus we have a one-to-one correspondence by which a finite sequence of natural numbers is associated with every finite sequence of primitive signs (hence also with every formula). We now map the finite sequences of natural numbers on natural numbers (again by a one-to-one correspondence), associating the number $2_i^n, 3_2^n, \dots, p_k^n k$, where p_k denotes the kth prime number (in order of increasing magnitude), with the sequence $n_i, n_2, \dots, n_k$. A natural number [out of a certain subset] is thus assigned one-to-one not only to every primitive sign but also to every finite sequence, of such signs. We denote by $\Phi(a)$ the number assigned to the primitive sign (or to the sequence of primitive signs) a. Now let some relation (or class) $R(a_i, a_2, \dots, a_n)$ between [or of] primitive signs or sequences of primitive signs be given. With it we associate the relation (or class) $R'(x_i, x_2, \dots, x_n)$ between [or of] natural numbers that obtains between $x_i, x_2, \dots, x_n$ if and only if there are some $a_i, a_2, \dots a_n$ such that $x_i = \Phi(a_i)$ $(i = 1, 2, \dots, n)$ and $R(a_i, a_2, \dots, a_n)$ hold. The relations between (or classes of) natural numbers that in this manner are associated with the metamathematical notions defined so far, for example. "variable", "formula", "sentential formula", "axiom", "provable formula". and so on, will be denoted by the same words in SMALL CAPITALS. The proposition that there are undecidable problems in the system P, for example, reads thus: There are SENTENTIAL FORMULAS a such that neither a nor the NEGATION of a is a PROVABLE FORMULA.

We now insert a parenthetic consideration that for the present has nothing to do with the formal system P. First we give the following definition: A number-theoretic function[29] $\varphi(x_i, x_2, \dots, x_n)$ is said to be *recursively defined*

---

[28] The rule of substitution is rendered superfluous by the fact that all possible substitutions have already been carried out in the axioms themselves. (This procedure was used also by *von Nemann 1927*.)

[29] That is, its domain of definition is the class of nonnegative integers (or of n-tuples of non-negative integers) and its values are nonnegative integers.

in terms of the number-theoretic functions $\psi(x_i, x_2, \ldots, x_{n-1})$ and $\mu(x_i, x_2, \ldots, x_{n+1})$ if

$$\varphi(0, x_2, \ldots, x_n) = \psi(x_2, \ldots, x_n),$$
$$\varphi(k + 1, x_2, \ldots, x_n) = \mu(k, \varphi(k, x_2, \ldots, x_n), x_2, \ldots, x_n) \qquad (3.2)$$

hold for all $x_2, \ldots, x_n, k$[30]

A number-theoretic function φ is said to be *recursive* if there is a finite sequence of number-theoretic functions $\varphi_i, \varphi_2, \ldots, \varphi_n$ that ends with φ and has the property that every function φ_k of the sequence is recursively defined in terms of two of the preceding functions, or results from any of the preceding functions by substitution.[31] or, finally, is a constant or the successor function $x + 1$. The length of the shortest-sequence of φ_i corresponding to a recursive function φ is called its *degree*. A relation $R(x_i, \ldots, x_n)$ between natural numbers is said to be recursive[32] if there is a recursive function $\varphi(x_i, \ldots, x_n)$ such that, for all $x_i, x_2, \ldots, x_n$,

$$R(x_i, \ldots, x_n) \sim [\varphi(x_i, \ldots, x_n) = 0].\text{[33]}$$

The following theorems hold:

I. *Every function (relation) obtained from recursive functions (relations) by substitution of recursive functions for the variables is recursive; so is every function obtained from recursive functions by recursive definition according to schema (2);*

II. *If B and S are recursive relations, so are $\overline{R}$ and $R \vee S$ (hence also $R \& S$);*

III. *If the functions $\varphi(\xi)$ and $\psi(\eta)$ are recursive, so is the relation $\varphi(\xi) = \psi(\eta)$;*[34]

---

[30] In what follows, lower-case italic letters (with or without subscripts) are always variables for nonnegative integers (unless the contrary is expressly noted).

[31] More precisely, by substitution of some of the preceding functions at the argument places of one of the preceding functions, for example, $\varphi_k(x_i, x_2) = \varphi_p[\varphi_q(x_i, x_2), \varphi_r(x_2)]$ $(p, q, r < k)$. Not all variables on the left side need occur on the right side (the same applies to the recursion schema (2)).

[32] We include classes among relations (as one-place relations). Recursive relations R, of course, have the property that for every given n-tuples of numbers it can be decided whether $R(x_i, \ldots, x_n)$ holds or not.

[33] Whenever formulas are used to express a meaning (in particular, in all formulas expressing meta-mathematical propositions or notions), Hilbert's symbolism is employed. See *Hilbert und Ackermann 1928*.

[34] We use German letters, ξ, η, as abbreviations for arbitrary n-tuples of variables, for example, $x_i, x_2, \ldots, x_n$.

IV. *If the function $\varphi(\xi)$ and the relation $R(x, \eta)$ are recursive, so are the relations S and T defined by*

$$S(\xi, \eta) \sim (Ex)[x \leqq \varphi(\xi) \& R(x, \eta)]$$

and

$$T(\xi, \eta) \sim (x)[x\varphi(\xi) \to R(x, \eta)],$$

as well as the function ψ defined by

$$\psi(\xi, \eta) = \varepsilon x[x\varphi(\xi) \& R(x, \eta)],$$

where $\varepsilon x F(x)$ means the least number x for which $F(x)$ holds and 0 in case there is no such number.

Theorem I follows at once from the definition of "recursive". Theorems II and III are consequences of the fact that the number-theoretic functions

$$\alpha(x), \beta(x, y), \gamma(x, y),$$

corresponding to the logical notions $\overline{}$, ν, and $=$, namely,

$$\alpha(0) = 1, \alpha(x) = 0 \text{ for } x \neq 0,$$
$$\beta(0, x) = \beta(x, 0) = 0, \beta(x, y) = 1 \text{ when } x \text{ and } y \text{ are both } \neq 0,$$
$$\gamma(x, y) = 0 \text{ when } x = y, \quad \gamma(x, y) = 1 \text{ when } x \neq y,$$

are recursive, as we can readily see. The proof of Theorem IV is briefly as follows. By assumption there is a recursive $\rho(x, \eta)$ such that

$$R(x, \eta) \sim [p(x, \eta) = 0].$$

We now define a function $\chi(x, \eta)$ by the recursion schema (2) in the following way:

$$\chi(0, \eta) = 0,$$
$$\chi(n + 1, \eta) = (n + 1).a + \chi(n, \eta).\alpha(a),\text{ }^{35}$$

where $a = \alpha[\alpha(\rho(0, \eta))].\alpha[\rho(n + 1, \eta)].\alpha[x(n, \eta)]$. Therefore $x(n + 1, \eta)$ is equal either to $n + 1$ (if $a = 1$) or to $x(n, \eta)$ (if $a = 0$).[36] The first case clearly occurs if and only if all factors of a are 1, that is, if

$$\overline{R}(0, \eta)) \& R(n + 1, \eta) \& [\chi(n, \eta) = 0]$$

---

[36] a cannot take values other than 0 and 1, as can be seen from the definition of α.

holds. From this it follows that the function $\chi(n, \eta)$ (considered as a function of n) remains 0 up to [but not including] the least value of n for which $R(n, \eta)$ holds and, from there on, is equal to that value. (Hence, in case $R(0, \eta)$ holds, $\chi(n, \eta)$ is constant and equal to 0.) We have, therefore,

$$\psi(\xi, \eta,) = \chi(\varphi(\xi), \eta),$$
$$S(\xi, \eta,) \sim R[\psi(\xi), \eta),$$

The relation T can, by negation, be reduced to a case analogous to that of S. Theorem IV is thus proved.

The functions $x + y$, $x \cdot y$, and x^y, as well as the relations $x < y$ and $x = y$, are recursive, as we can readily see. Starting from these notions, we now define a number of functions (relations) 1–45, each of which is defined in terms of preceding ones by the procedures given in Theorems I–IV. In most of these definitions several of the steps allowed by Theorems I–IV are condensed into one. Each of the functions (relations) 1–45, among which occur, for example, the notions "FORMULA", "AXIOM", and "IMMEDIATE CONSEQUENCE", is therefore recursive.

1. $x/y \equiv (Ez)[z \leq \& x = y.z]$,[37]
   x is divisible by y.[38]
2. $\text{Prim}(x) \equiv \overline{(Ex)}[z \leq x \& z \neq 1 \& z \neq x \& x/z] \& x > 1$,
   x is a prime number.
3. $0 \, Pr \, x \equiv 0$,
   $(n + 1) Pr \, x \equiv \varepsilon y[y \leq x \& \text{Prim}(y) \& x/y \& y > n \, Pr \, x]$,
   $n \, Pr \, x$ is the nth prime number (in order of increasing magnitude) contained in x. For $0 < n \leq z$, where z is the number of distinct prime factors of x. Note that $n \, Pr \, x = 0$ for $n = z + 1$.
4. $0! \equiv 1$,
   $(n + 1)! \equiv (n + 1).n!$.
5. $Pr(0) \equiv 0$,
   $Pr(n \div 1) \equiv \varepsilon y[y \leq \{Pr(n)\}! + 1 \& \text{Prim}(y) \& y > Pr(n)]$,
   $Pr(n)$ is the nth prime number (in order of increasing magnitude).
6. $n \, Gl \, x \equiv \varepsilon y[y \leq x \& x/(n \, Pr \, x)^y \& \overline{x(n \, Pr \, x)}^{y+1}]$,

---

[37] The sign $\equiv$ is used in the sense of "equality by definition"; hence in definitions it stands for either $=$ or $\sim$ (otherwise, the symbolism is Hilbert's).

[38] Wherever one of the signs (x), (Ex), or Ex occurs in the definitions below, it is followed by a bound on x. This bound merely serves to ensure that the notion defined is recursive (see Theorem IV). But in most cases the *extension* of the notion defined would not change if this bound were omitted.

$n\ Gl\ x$ is the nth term of the number sequence assigned to the number x (for $n > 0$ and n not greater than the length of this sequence).

7. $l(x) \equiv \varepsilon y[y \leq x\ \&\ y\ Pr\ x > 0\ \&\ (y \div 1)Pr\ x = 0]$,

   $l(x)$ is the length of the number sequence assigned to x.

8. $x * y \equiv \varepsilon z\{z \leq [Pr(l(x) + 1(y))]^{x+y}\ \&\ (n)[n \leq l(x) \rightarrow n\ Gl\ z = n\ Gl\ x]\ \&\ (n)[0 < n \leq l(y) \rightarrow (n + l(x))Gl\ z = n\ Gl\ y]\}$,

   $x * y$ corresponds to the operation of "concatenating" two finite number sequences.

9. $R(x) \equiv 2^x$,

   $R(x)$ corresponds to the number sequence consisting of x alone (for $x > 0$).

10. $E(x) \equiv R(11) * x * R(13)$,

    $E(x)$ corresponds to the operation of "enclosing within parentheses" (11 and 13 are assigned to the primitive signs "(" and ")", respectively).

11. $n\ Var\ x \equiv (Ez)[13 < z \leq x\ \&\ \text{Prim}(z)\ \&\ x = z^n]\ \&\ n \neq 0$,

    x is a VARIABLE OF TYPE n.

12. $\text{Var}(x) \equiv (En)[n \leq x\ \&\ n\ Var\ x]$,

    x is a VARIABLE.

13. $\text{Neg}(x) \equiv R(5) * E(x)$,

    $\text{Neg}(x)$ is the NEGATION of x.

14. $x\ Dis\ y \equiv E(x) * R(7) * E(y)$,

    $x\ Dis\ y$ is the DISJUNCTION of x and y.

15. $x\ Gen\ y \equiv R(x) * R(9) * E(y)$,

    $x\ Gen\ y$ is the GENERALIZATION of y with respect to the VARIABLE x (provided x is a VARIABLE).

16. $0\ Nx \equiv x$,

    $(n + 1)Nx \equiv R(3) * nNx$,

    nNx corresponds to the operation of "putting the sign 'f' n times in front of x".

17. $Z(n) \equiv n\ N[R(1)]$,

    $Z(n)$ is the NUMERAL denoting the number n.

18. $\text{Typ}'_1(x) \equiv (Em, n)\{m, n \leq x\ \&\ [m = 1 \lor 1\ Var\ m]\ \&\ x = n\ N[R(m)]\}$[39]

    x is a SIGN OF TYPE 1.

19. $\text{Typ}_n(x) \equiv [n = 1\ \&\ \text{Typ}'_1(x)]\lor[n > 1\ \&\ (Ev)\{v \leq x\ \&\ n\ Var\ v\ \&\ x = R(v)\}]$,

    x is a SIGN OF TYPE n.

---

[39] $m, n \leq x$ stands for $m \leqq x\ \&\ n \leqq x$ (similarly for more than two variables).

20. $Elf(x) \equiv (Ey, z, n)[y, z, n \leq x \& \text{Typ}_x(y) \& \text{Typ}_{n+1}(z) \& x = z * E(y)]$,

   x is an ELEMENTARY FORMULA.

21. $Op(x, y, z) \equiv x = \text{Neg}(y) vx = y \text{ Dis } zv(Ev)[v \leq x \& \text{Var}(v) \& x = v \text{ Gen } y]$.

22. $FR(x) \equiv (n)\{0n \leq l(x) \rightarrow Elf(nGlx) \vee (Ep, q)[0 < p, q < n \& Op(n Glx, p Glx, q Glx)]\} \& l(x) > 0$,

   x is a SEQUENCE OF FORMULAS, each of which either is an ELEMENTARY FORMULA or results from the preceding FORMULAS through the operations of NEGATION, DISJUNCTION, or GENERALIZATION.

23. $\text{Form}(x) \equiv (En)\{n \leq (Pr[l(x)^2])^{x-[l(x)]^2} \& FR(n) \& x = [l(n)] Gl n\}$,[40]

   x is a FORMULA (that is, the last term of a FORMULA SEQUENCE n).

24. $v \text{ Geb } n, x \equiv \text{Var}(v) \& \text{Form}(x) \& Ea, b, c)[a, b, c \leq x \& x = a * (v\text{Gen}b) * c \& \text{Form}(b) \& l(a) + 1 \leq n \leq l(a) + l(v \text{ Gen } b)]$,

   the VARIABLE v is BOUND in x at the nth place.

25. $v \text{ Fr } n, x \equiv \text{Var}(v) \& \text{Form}(x) \& v = n Gl x \& n \leq l(x) \& \overline{v \text{ Geb } n, x}$,

   the VARIABLE v is FREEin x at the nth place.

26. $v \text{ Fr } x \equiv (En)[n \leq l(x) \& v \text{ Fr } n, x]$,

   v occurs as a FREE VARIABLE in x.

27. $Su \, x\binom{n}{y} \equiv \varepsilon z\{z \leq [Pr(l(x) + l(y))]^{x+y} \& [(Eu, v)u, v x \& x = u * R(n \leq Gl x) * v \& z = u * y * v \& n = l(u) + 1]\}$,

   $Su \, x\binom{n}{y}$ results from x when we substitute y for the nth term of x (provided that $0 < n \leq l(x)$).

28. $0 \, St \, v, x \equiv \varepsilon n\{n \leq l(x) \& v \text{ Fr } n, x \& \overline{(E_p)}[n < p \leq l(x) \& v \text{ Fr } p, x]\}, (k + 1)Stv, x \equiv \varepsilon n\{n < k \, St \, v, x \& r \text{ Fr } n, x \& \overline{(E_p)}[n < p < k \, St \, v, x \& v \text{ Fr } p, x]\}$,

   $k \, St \, v, x$ is the $(k + 1)$th place in x (counted from the right end of the FORMULA x) at which v is FREE x (and 0 in case there is no such place).

29. $A(v, x) \equiv \varepsilon n\{n \leq l(x) \& n \, St \, v, x = 0\}$,

   $A(v, x)$ is the number of places at which v is FREE in x.

---

[40] That $n(Pr \leq [l(x)]^2)^{(x).[l(x)]^2}$ provides a bound can be seen thus: The length of the shortest sequence of formulas that corresponds to x can at most be equal to the number of subformulas of x. But there are at most $l(x)$ subformulas of length 1, at most $l(x) - 1$ of length 2, and so on, hence altogether at most $l(x)(l(x) + 1)/2 \leq [l(x)]^2$. Therefore all prime factors of n can be assumed to be less than $Pr([l(x)]^2)$, their number $\leq [(lx)]^2$, and their exponents (which are subformulas of x) $\leq x$.

30. $Sb_0(x_y^v) \equiv x,$

$$Sb_{k+1}(x_y^v) \equiv Su[Sb_k(x_y^v)] \binom{k \, St \, v, \, x}{y}.$$

31. $Sb(x_y^v) \equiv Sb_{A(v,x)}(x_y^v),$ [41]

$Sb(x_y^v)$ is the notion SUBST $a \binom{v}{b}$ defined above. [42]

32. $x \, \text{Imp} \, y \equiv [\text{Neg}(x)] \, \text{Dis} \, y,$
    $x \, \text{Con} \, y \equiv \text{Neg}\{[\text{Neg}(x)] \, \text{Dis} \, [\text{Neg}(y)]\},$
    $x \, \text{Aeq} \, y \equiv (x \, \text{Imp} \, y) \, \text{Con} \, (y \, \text{Imp} \, x),$
    $v \, \text{Ex} \, y \equiv \text{Neg}\{v \, \text{Gen} \, [\text{Neg}(y)]\}.$

33. $n \, Th \, x \equiv \varepsilon y\{y \leq x^{(2z)} \, \& \, (k)[k \leq l(x) \rightarrow (k \, Gl \, x \leq 13 \, \& \, k \, Gl \, y = k \, Gl \, x) \vee (k \, Gl \, x > 13 \, \& \, k \, Gl \, y = k \, Gl \, x.[1 \, Pr(k \, Gl \, x)]^n)]\},$

    $n \, Th \, x$ is the nth TYPE ELEVATION of x (in case x and $n \, Th \, x$ are FORMULAS).

    Three specific numbers, which we denote by z_i, z_2 and z_3, correspond to the Axioms I, 1-3, and we define

34. $Z - Ax(x) \equiv (x = z_1 \vee x = z_2 \vee x = z_3).$

35. $A_i - Ax(x) \equiv (Ey)[y \leq x \, \& \, \text{Form}(y) \, \& \, x = (y \, \text{Dis} \, y) \, \text{Imp} \, y],$

    x is a FORMULA resulting from Axiom schema II, 1 by substitution. Analogously, $A_2 - Ax$, $A_3 - Ax$, and $A_4 - Ax$ are defined for Axioms [rather, Axiom Schemata] II, 2-4.

36. $A - Ax(x) \equiv A_1 - Ax(x) \vee A_2 - Ax(x) \vee A_3 - Ax(x) \vee A_4 - Ax(x),$

    x is a FORMULA resulting from a propositional axiom by substitution.

37. $Q(z, y, r) \equiv \overline{(En, m, u)}[n \leq l(y) \, \& \, m \leq l(z) \& w \leq z \& w \equiv m \, Gl \, z \, \& \, w \, \text{Ge} \, b \, n, \, y \& v \, Fr \, n, y]$

    z does not contain any VARIABLE BOUND in y at a place at which v is FREE.

38. $L_i - Ax(x) \equiv (Ev, y, z, n)\{v, y, z, n \leq x \, \& \, n \, \text{Var} \, v \& \text{Typ}_n(z) \& \text{Form}(y) \, \& \, Q(z, y, v) \, \& \, x = (v \, \text{Gen} \, y) \, \text{Imp} \, [Sb(y_z^v)]\},$

    x is a FORMULA resulting from Axiom schema III, 1 by substitution.

39. $L_2 - Ax(x) \equiv (Ev, q, p)\{v, q, p \leq x \, \& \, \text{Var}(v) \, \& \, \text{Form}(p) \& \overline{v \, Fr \, p} \& \text{Form}(q) \, \& \, x = [v \, \text{Gen} \, (p \, \text{Dis} \, q)] \, \text{Imp} \, [p \, \text{Dis} \, (v \, \text{Gen} \, q)]\},$

    x is a FORMULA resulting from Axiom schema III, 2 by substitution.

40. $R - Ax(x) \equiv (Eu, v, y, n)[u, v, y, n \leq x \, \& \, n \, \text{Var} \, v \, \& \, (n + 1) \, \text{Var} \, u \& \overline{u \, Fr \, y} \& \text{Form}(y) \& x = u \, Ex\{v \, \text{Gen} \, [[R(u) * E(R(v))] \, \text{Aeq} \, y]\}],$

---

[41] In case v is not a VARIABLE or x is not a FORMULA, $Sb(x_y^v) = x$.

[42] Instead of $Sb(Sb(x_y^v)_z^w)$ we write $Sb(x_{yz}^{vw})$ (and similarly for more than two VARIABLES).

x is a FORMULA resulting from Axiom schema IV, 1 by substitution.

A specific number z_4 corresponds to Axiom V, 1, and we define:

41. $M - Ax(x) \equiv (En)[n \leq x \ \& \ x = n \, Th \, z_4]$.

42. $Ax(x) \equiv Z - Ax \vee (x)A - Ax(x)vL_i - Ax(x)vL_2 - Ax(x)vR - Ax(x)v$
    $M - Ax(x)$,

    x is an AXIOM.

43. $Fl(x, y, z) \equiv y = z \, Imp \, xv(Ev)[v \leq x \ \& \ Var(v) \& x = v \, Gen \, y]$,

    x is an IMMEDIATE CONSEQUENCE of y and z.

44. $Bw(x) \equiv (n)\{0 < n \leq l(x) \rightarrow Ax(n \, Gl \, x)v(Ep, q)[0 < p, q < n$
    $\& Fl(n \, Gl \, x, p \, Gl \, x, q \, Gl \, x)]\} \ \& \ l(x) > 0$,

    x is a PROOF ARRAY (a finite sequence of FORMULAS, each of which is either an AXIOM or an IMMEDIATE CONSEQUENCE of two of the preceding FORMULAS.

45. $x \, B \, y \equiv Bw(x) \ \& \ [l(x)] \, Gl \, x = y$,

    x is a PROOF OF THE FORMULA y.

46. $Bew(x) \equiv (Ey)y \, B \, x$,

    x is a PROVABLE FORMULA. ($Bew(x)$, is the only one of the notions 1–46 of which we cannot assert that it is recursive.)

The fact that can be formulated vaguely by saying: every recursive relation is definable in the system P (if the usual meaning is given to the formulas of this system), is expressed in precise language, *without* reference to any interpretation of the formulas of P, by the following theorem:

Theorem I. *For every recursive relation $R(x_i, \ldots, x_n)$ there exists an n-place RELATION SIGN r (with the FREE VARIABLES*[43] *$u_1, u_2, \ldots, u_n$) such that for all n-tuples of numbers $(x_i, \ldots, x_n)$ we have*

$$R(x_i, \ldots, x_n) \rightarrow Bew[Sb(r^{x_i - x_n}_{Z(x_i) - Z(x_n)})] \tag{3.3}$$

$$\overline{R}(x_i, \ldots, x_n) \rightarrow Bew[Neg(Sb(r^{x_i \quad x_n}_{Z(x_i) - Z(x_n)}))] \tag{3.4}$$

We shall give only an outline of the proof of this theorem because the proof does not present any difficulty in principle and is rather long.[44] We prove the theorem for all relations $R(x_i, \ldots, x_n)$ of the form $x_i = \varphi(x_2, \ldots, x_n)$[45]

---

[43] The VARIABLES $u_i, \ldots, u_n$ can be chosen arbitrarily. For example, there always is an r with the FREE VARIABLES 17, 19, 23, …, and so on for which (3) and (4) hold.

[44] Theorem V, of course, is a consequence of the fact that in the case of a recursive relation R it can, for every n-tuple of numbers, be decided *on the basis of the axioms of the system P* whether the relation R obtains or not.

[45] From this it follows at once that the theorem holds for every recursive relation, since any such relation is equivalent to $0 = \varphi(x_i, \ldots, x_n)$, where φ is recursive.

(where φ is a recursive function) and we use induction on the degree of φ. For functions of degree 1 (that is, constants and the function $x + 1$) the theorem is trivial. Assume now that φ is of degree m. It results from functions of lower degrees, $\varphi_i, \ldots, \varphi_k$, through the operations of substitution or recursive definition. Since by the induction hypothesis everything has already been proved for $\varphi_i, \ldots, \varphi_k$, there are corresponding RELATION SIGNS, $r_i, \ldots, r_k$ such that (3) and (4) hold. The processes of definition by which φ, results from $\varphi_i, \ldots, \varphi_k$, (substitution and recursive definition) can both be formally reproduced in the system P. If this is done, a new RELATION SIGN r is obtained from $r_i, \ldots, r_k,$[41] and, using the induction hypothesis, we can prove without difficulty that (3) and (4) hold for it. A RELATION SIGN r assigned to a recursive relation[46] by this procedure will be said to be recursive.

We now come to the goal of our discussions. Let k be any class of FOR-MULAS. We denote by $\mathrm{Flg}(k)$ (the set of consequences of k) the smallest set of FORMULAS that contains all FORMULAS of k and all AXIOMS and is closed under the relation "IMMEDIATE CONSEQUENCE". k is said to be w-consistent if there is no CLASS SIGN a such that

$$(n)[Sb(a^v_{z(n)})\varepsilon\,\mathrm{Flg}(k)] \,\&\, [Neg(v\,\mathrm{Gen}\,a)]\varepsilon\,\mathrm{Flg}(k),$$

where v is the FREE VARIABLE of the CLASS SIGN a.

Every w-consistent system, of course, is consistent. As will be shown later, however, the converse does not hold.

The general result about the existence of undecidable propositions reads as follows:

Theorem II. *For every w-consistent recursive class k of* FORMULAS *there are recursive* CLASS SIGNS *r such that neither v Gen r nor Neg($\vee$ Gen r) belongs to* $\mathrm{Flg}(k)$ *(where v is the* FREE VARIABLE *of r).*

Proof. Let k be any recursive w-consistent class of FORMULAS. We define

$$Bw_k(x) \equiv (n)[n \leq l(x) \rightarrow Ax(n\,Gl\,x)v(n\,Gl\,x)\varepsilon\,k\,v$$
$$(Ep,q)\{0 < p,q < n \,\&\, Fl(n\,Gl\,x, p\,Gl\,x, q\,Gl\,x)\}]\&, l(x) > 0 \quad (3.5)$$

(see the analogous notion 44),

$$x\,B_k y \equiv Bw_k(x)\&[l(x)]\,Gl\,x = \qquad (3.6)$$

---

[46] When this proof is carried out in detail, r, of course is not defined indirectly with the help of its meaning but in terms of its purely formal structure.

$$Bew_k(x) \equiv (Ey)y \, B_k x \tag{6.1}$$

(see the analogous notions 45 and 46). $\qquad\square$

We obviously have

$$(x)[Bew_k(x) \sim x \, \varepsilon \, Flg(k)] \tag{3.8}$$

and

$$(x)[Bew(x) \to Bew_k(x)]. \tag{3.9}$$

We now define the relation

$$Q(x, y) \equiv \overline{x \, B_k[Sb(y_{z(y)}^{19})]}. \tag{3.10}$$

Since $x \, B_k \, y$ (by (6) and (5)) and $Sb(y_{z(y)}^{19})$ (by Definitions 17 and 31) are recursive, so is $Q(x, y)$. Therefore, by Theorem V and (8) there is a RELATION SIGN q (with the FREE VARIABLES 17 and 19) such that

$$\overline{x \, B_k[Sb(y_{z(y)}^{19})]} \to Bew_k[Sb(q_{z(x)z(y)}^{17 \ 19})], \tag{3.11}$$

and

$$\overline{x \, B_k[Sb(y_{z(y)}^{19})]} \to Bew_k[Neg(Sb(q_{z(x)z(y)}^{17 \ 19}))]. \tag{3.12}$$

We put

$$p = 17 \, Gen \, q \tag{3.13}$$

(p is a CLASS SIGN with the FREE VARIABLE 19) and

$$r = Sb(q_{z(p)}^{19}) \tag{3.14}$$

(r is a recursive CLASS SIGN[47] with the FREE VARIABLE 17).

Then we have

$$Sb(p_{z(p)}^{19}) = Sb([17 \, Gen \, q]_{z(p)}^{19}) = 17 \, Gen \, Sb(q_{z(p)}^{19}) = 17 \, Gen \, r \tag{3.15}$$

(by (11) and (12));[48] furthermore

$$Sb(q_{z \ z(p)}^{17 \, 19}) = Sb(r_{z(x)}^{17}) \tag{3.16}$$

---

[47] Since r is obtained from the recursive RELATION SIGN q through the replacement of a VARIABLE by a definite number, p. [Precisely stated the final part of this footnote (which refers to a side remark unnecessary for the proof) would read thus: "REPLACEMENT of a VARIABLE: by the NUMERAL for p."]

[48] The operations Gen and Sb, of course, can always be interchanged in case they refer to different VARIABLES.

(by (12)). If we now substitute p for y in (9) and (10) and take (13) and (14) into account, we obtain

$$\overline{x \, B_k(17 \, Gen \, r)} \to Bew_k[Sb(r_{z(x)}^{17})], \tag{3.17}$$

$$x \, B_k(17 \, Gen \, r) \to Bew_k[Neg(Sb(r_{z(x)}^{17}))]. \tag{3.18}$$

This yields:

1. 17 Gen r is not k-PROVABLE.[49] For, if it were, there would (by (6.1)) be an n such that $n \, B_\kappa (17 \, Gen \, r)$. Hence by (16) we would have $Bew_k[Neg (Sb(r_{z(n)}^{17}))]$, while, on the other hand, from the κ-PROVABILITY of 17 Gen r that of $Sb(r_{z(n)}^{17})$ follows. Hence, κ would be inconsistent (and a fortiori ω-inconsistent).

2. Neg(17 Gen r) is not κ-PROVABLE. Proof: As has just been proved, 17 Gen r is not κ-PROVABLE; that is (by (6.1)), $(n)\overline{n \, B_k \, (17 \, Gen \, r)}$ holds. From this, $(n)\overline{Bew_k[Sb(r_{z(n)}^{17})]}$ follows by (15), and that, in conjunction with $Bew_k[Neg(17 \, Gen \, r)]$, is incompatible with the ω-consistency of κ. 17 Gen r is therefore undecidable on the basis of κ, which proves Theorem VI.

   We can readily see that the proof just given is constructive;[50] that is, the following has been proved in an intuitionistically unobjectionable manner: Let an arbitrary recursively defined class κ of FORMULAS be given. Then, if a formal decision (on the basis of κ) of the SENTENTIAL FORMULA 17 Gen r (which [for each κ] can actually be exhibited) is presented to us, we can actually give

   (a) A PROOF of Neg(17 Gen r);
   (b) For any given n, a PROOF of $Sb(r_{z(n)}^{17})$.

   That is, a formal decision of 17 Gen r would have the consequence that we could actually exhibit an ω-inconsistency.

We shall say that a relation between (or a class of) natural numbers $P(x_i, \ldots, x_n)$ is [[*entscheidungsdefinit*]] if there exists an n-place RELATION SIGN r such that (3) and (4) (see Theorem V) hold. In particular, therefore, by Theorem V every recursive relation is decidable. Similarly, a RELATION SIGN will be said to be *decidable* if it corresponds in this way to a decidable relation. Now it suffices for the existence of undecidable propositions that the

---

[49] By "x is *k-provable*" we mean $x \, \varepsilon \, Flg_{(k)}$, which, by (7), means the same thing as $Bew_K(x)$.

[50] Since all existential statements occurring in the proof are based upon Theorem V. which. as is easily seen, is unobjectionable from the intuitionistic point of view.

class κ be ω-consistent and decidable. For the decidability carries over from κ to $x B_\kappa y$ (see (5) and (6)) and to $Q(x, y)$ (see (8.1)), and only this was used in the proof given above. In this case the undecidable proposition has the form υ Gen r, where r is a decidable CLASS SIGN. (Note that it even suffices that κ be decidable in the system enlarged by κ.)

If, instead of assuming that κ is ω-consistent, we assume only that it is consistent, then, although the existence of an undecidable proposition does not follow [by the argument given above]. it does follow that there exists a property (r) for which it is possible neither to give a counterexample nor to prove that it holds of all numbers. For in the proof that 17 Gen r is not κ-PROVABLE only the consistency of k was used (see p. 608). Moreover from $\overline{\text{Bew}}_k(17\,\text{Gen}\,r)$ it follows by (15) that, for every number x, $Sb(r^{17}_{z(x)})$ is κ-PROVABLE and consequently that $\text{Neg}(Sb(r^{17}_{z(x)}))$ is not k-PROVABLE: for any number.

If we adjoin $\text{Neg}(17\,\text{Gen}\,r)$ to κ, we obtain a class of FORMULAS $k\prime$ that is consistent but not ω-consistent, $k\prime$ is consistent, since otherwise 17 Gen r would be κ-PROVABLE. However, $k\prime$ is not ω-consistent, because, by $\overline{\text{Bew}}_k(17\,\text{Gen}\,r)$ and (15), $(x)\text{Bew}_k Sb(r^{17}_{z(x)})$ and, a fortiori, $(x)\text{Bew}_k St(r^{17}_{z(x)})$ hold, while on the other hand, of course, $\text{Bew}_k[\text{Neg}(17\,\text{Gen}\,r)]$ holds.[51]

We have a special case of Theorem VI when the class k consists of a finite number of FORMULAS (and, if we so desire, of those resulting from them by TYPE ELEVATION).

Every finite class κ is, of course, recursive.[52] Let a be the greatest number contained in κ. Then we have for κ

$$x \varepsilon \kappa \sim (Em, n)[m \leqq x \;\&\; n \leqq a \;\&\; n \varepsilon \kappa \;\&\; x = m \, Th \, n].$$

Hence κ is recursive. This allows us to conclude, for example, that, even with the help of the axiom of choice (for all types) or the generalized contin uum hypothesis, not all propositions are decidable, provided these hypotheses are ω-consistent.

In the proof of Theorem VI no properties of the system P were used besides the following:

---

[51] Of course, the existence of classes κ that are consistent but not ω-consistent is thus proved only on the assumption that there exists some consistent κ (that is, that P is consistent).

[52] [[On page 190, lines 21, 22, and 23, of the German text the three occurrences of a are misprints and should be replaced by occurrences of k.]]

1. The class of axioms and the rules of inference (that is, the relation "immediate consequence") are recursively definable (as soon as we replace the primitive signs in some way by natural numbers);
2. Every recursive relation is definable (in the sense of Theorem V) in the system P. Therefore, in every formal system that satisfies the assumptions 1 and 2 and is ω-consistent there are undecidable propositions of the form $(x)F(x)$, where F is a recursively defined property of natural numbers, and likewise in every extension of such a system by a recursively definable ω-consistent class of axioms. As can easily be verified, included among the systems satisfying the assumptions 1 and 2 are the Zermelo-Fraenkel and the von Neumann axiom systems of set theory,[53] as well as the axiom system of number theory consisting of the Peano axioms, recursive definition (by schema (2)), and the rules of logic.[54] Assumption 1 is satisfied by any system that has the usual rules of inference and whose axioms (like those of P) result from a finite number of schemata by substitution.[55]

We shall now deduce some consequences from Theorem VI, and to this end we give the following definition:

A relation (class) is said *to be arithmetic* if it can be defined in terms of the notions + and . (addition and multiplication for natural numbers)[56] and the logical constants $\vee$, $^{\frown}$, (x), and =, where (x) and = apply to natural numbers only.[57] The notion "arithmetic proposition" is defined accordingly. The relations "greater than" and "congruent modulo n", for example, are

---

[53] The proof of assumption 1 turns out to be even simpler here than for the system P, since there is just one. kind of primitive variables (or two in von Neumann's system).

[54] See Problem III in *Hilbert 1928a*.

[55] As will be shown in Part II of this paper, the true reason for the incompleteness inherent in all formal systems of mathematics is that the formation of ever higher types can be continued into the transfinite (see *Hilbert 1925*, p. 184 [[above, p. 387]), while in any formal system at most denumerably many of them are available. For it can be shown that the undecidable propositions constructed here become decidable whenever appropriate higher types are added (for example, the type w to the system P). An analogous situation prevails for the axiom system of set theory.

[56] Here and in what follows, zero is always included among the natural numbers.

[57] The definiens of such a notion, therefore, must consist exclusively of the signs listed, variables for natural numbers, x, y, . . . , and the signs 0 and 1 (variables for functions and sets are not permitted to occur). Instead of x any other number variable, of course, may occur in the prefixes.

arithmetic because we have

$$x > y \sim \overline{(Ez)}[y = x + z],$$
$$x \equiv y(\bmod n) \sim (Ez)[x = y + z.n \vee y = x + z.n].$$

We now have

Theorem III. *Every recursive relation is arithmetic.*

We shall prove the following version of this theorem: every relation of the form $x_0 = \varphi(x_i, \ldots, x_n)$, where φ is recursive, is arithmetic, and we shall use induction on the degree of φ. Let φ be of degree $s(s > 1)$. Then we have either

1. $\varphi(x_i, \ldots, x_n) \quad = \quad \rho[\chi_i(x_i, \ldots, x_n), \chi_2(x_i, \ldots, x_n), \ldots, \chi_m(x_i, \ldots, x_n)]^{58}$ (where ρ and all χ_i are of degrees less than s) or
2. $\varphi(0, x_2, \ldots, x_n) = \psi(x_2, \ldots, x_n)$,

$$\varphi(k + 1, x_2, \ldots, x_n) = \mu[k, \varphi(k, x_2, \ldots, x_n), x_2, \ldots, x_n]$$

(where ψ and μ are of degrees less than s).

In the first case we have

$$x_0 = \varphi(x_i, \ldots, x_n) \sim (Ey_i, \ldots, y_m)[R(x_0, y_i, \ldots, y_m) \,\&$$
$$S_i(y_i, x_i, \ldots, x_n) \,\& \, \ldots \,\& \, S_m(y_m, x_i, \ldots, x_n)],$$

where R and S_i are the arithmetic relations, existing by the induction hypothesis, that are equivalent to $x_0 = \rho(y_i, \ldots, y_m)$ and $y = \chi_i(x_i, \ldots, x_n)$, respectively. Hence in this case $x_0 = \phi(x_i, \ldots, x_n)$ is arithmetic.

In the second case we use the following method. We can express the relation $x_0 = \phi(x_i, \ldots, x_n)$ with the help of the notion "sequence of numbers" $(f)^{59}$ in the following way:

$$x_0 = \varphi(x_1, \ldots, x_n) \sim (Ef)\{f_0 = \psi(x_2, \ldots, x_n) \& (k)[k < x_i$$
$$\rightarrow f_{k+1} = \mu(k, f_k, x_2, \ldots, x_n)] \& x_0 = f_{x_1}\}.$$

If $S(y, x_2, \ldots, x_n)$ and $T(z, x_i, \ldots, x_{n+1})$ are the arithmetic relations, existing by the induction hypothesis, that are equivalent to $y = \psi(x_2, \ldots, x_n)$

---

[58] Of course, not all $x_1, \ldots, x_n$ need occur in the x_i (see the example in footnote 27).

[59] f here is a variable with the [infinite] sequences of natural numbers as its domain of values, f_k denotes the $(k + 1)$th terra of a sequence f (f_0 denoting the first).

and $z = \mu(x_i, \ldots, x_{n+1})$, respectively, then

$$x_0 = \varphi(x_i, \ldots, x_n) \sim (Ef)\{S(f_0, x_2, \ldots, x_n)\&(k)[k < x_2$$
$$\rightarrow T(f_{k+1}, k, f_k, x_2, \ldots, x_n)]\&x_0 = f_{x1}\}. \tag{3.19}$$

We now replace the notion "sequence of numbers" by "pair of numbers", assigning to the number pair n, d the number sequence $f^{(n,d)}(f_k^{(n,d)} = [n]_{1+(k+1)d})$, where $[n]_p$ denotes the least nonnegative remainder of n modulo p.

We then have

Lemma 1. If f is any sequence of natural numbers and k any natural number, there exists a pair of natural numbers, n, d such that $f^{(x,d)}$ and f agree in the first k terms.

Proof. Let l be the maximum of the numbers k, $f_0, \ldots, f_{k-1}$. Let us determine an n such that

$$n \equiv f_i[\mathrm{mod}(1 + (i + 1)l!)] \text{ for } i = 0, 1, \ldots, k - 1,$$

which is possible, since any two of the numbers $1 + f(i \div 1)l! \ (i = 0, 1 \ldots, k - 1)$ are relatively prime. For a prime number contained in two of these numbers would also be contained in the difference $(i_1 - i_2)l!$ and therefore, since $|i_i - i_2| < l$, in $l!$; but this is impossible. The number pair $n, l!$ then has the desired property.   □

Since the relation $x = [n]_p$ is defined by

$$x \equiv n(\mathrm{mod}\ p) \ \& \ x < p$$

and is therefore arithmetic, the relation $P(x_0, x_i, \ldots, x_n)$, defined as follows:

$$P(x_0, \ldots, x_n) \equiv (En, d)\{S([n]_{d \div 1}, x_2, \ldots, x_n) \ \& \ (k)[k < x_i$$
$$\rightarrow T([n]_{1 \div d(k+2)}, k, [n]_{1-d(k+1)}, x_2, \ldots, x_n)] \ \& \ x_0 = [n]_{1+d(x_1 \div 1)}\},$$

is also arithmetic. But by (17) and Lemma 1 it is equivalent to $x_0 = \varphi(x_i, \ldots, x_n)$ (the sequence f enters in (17) only through its first $x_i + 1$ terms). Theorem VII is thus proved.

By Theorem VII, for every problem of the form $(x)F(x)$ (with recursive F) there is an equivalent arithmetic problem. Moreover, since the entire proof of Theorem VII (for every particular F) can be formalized in the system P, this equivalence is provable in P. Hence we have

Theorem IV. *In any of the formal systems mentioned in Theorem VI[60] there are undecidable arithmetic propositions.*

By the remark on page 610, the same holds for the axiom system of set theory and its extensions by ω-consistent recursive classes of axioms.

Finally, we derive the following result:

Theorem V. *In any of the formal systems mentioned in Theorem VI[53] there are undecidable problems of the restricted functional calculus[61]* (that is, formulas of the restricted functional calculus for which neither validity nor the existence of a counterexample is provable).[62]

This is a consequence of

Theorem VI. *Every problem of the form* $(x)F(x)$ *(with recursive F) can be reduced to the question whether a certain formula of the restricted functional calculus is satisfiable* (that is, for every recursive F we can find a formula of the restricted functional calculus that is satisfiable if and only if $(x)F(x)$ is true.

By formulas of the restricted functional calculus (r. f. c.) we understand expressions formed from the primitive signs $\overline{}$, v, (x), $=$, $x, y, \ldots$ (individual variables), $F(x)$, $G(x, y)$, $H(x, y, z)$, ... (predicate and relation variables), where (x) and $=$ apply to individuals only.[63] To these signs we add a third kind of variables, $\varphi(x)$, $\psi(x, y)$, $\kappa(x, y, z)$, and so on, which stand for object-functions [[Gegenstandsfunktionen]] (that is, $\varphi(x)$, $\psi(x, y)$, and so on denote single-valued functions whose arguments and values are individuals).[64] A formula that contains variables of the third kind in addition

---

[60] These are the ω-consistent systems that result from P when recursively definable classes of axioms are added.

[61] See *Hilbert and Ackermann 1928*.

In the system P we must understand by formulas of the restricted functional calculus those that result from the formulas of the restricted functional calculus of *PM* when relations are replaced by classes of higher types as indicated on page 599.

[62] In *1930a* I showed that every formula of the restricted functional calculus either can be proved to be valid or has a counterexample. However, by Theorem IX the existence of this counterexample is *not* always provable (in the formal systems we have been considering).

[63] Hilbert and Ackermann (*1928*) do not include the sign $=$ in the restricted functional calculus. But for every formula in which the sign $=$ occurs there exists a formula that does not contain this sign and is satisfiable if and only if the original formula is (see *Gödel 1930a*).

[64] Moreover, the domain of definition is always supposed to be the *entire* domain of individuals.

to the signs of the r. f. c. first mentioned will be called a formula in the extended sense (i. e. s.).[65] The notions "satisfiable" and " valid" carry over immediately to formulas i.e. s., and we have the theorem that, for any formula A i.e. s., we can find a formula B of the r.f.c. proper such that A is satisfiable if and only if B is. We obtain B from A by replacing the variables of the third kind, $\varphi(x)$, $\psi(x, y)$, . . ., that occur in A with expressions of the form $(Iz)F(z, x)$, $(Iz)G(z, x, y)$, . . ., by eliminating the "descriptive" functions by the method used in *PM* (I, *14), and by logically multiplying[66] the formula thus obtained by an expression stating about each F, G, . . . put in place of some φ, ψ, . . . that it holds for a unique value of the first argument [for any choice of values for the other arguments].

We now show that, for every problem of the form $(x)F(x)$ (with recursive F), there is an equivalent problem concerning the satisfiability of a formula i. e. s., so that, on account of the remark just made, Theorem X follows.

Since F is recursive, there is a recursive function $\Phi(x)$ such that $F(x) \sim [\Phi(x) = 0]$, and for Φ there is sequence of functions, Φ_i, Φ_2, . . . , Φ_n, such that $\Phi_n = \Phi$, $\Phi_i(x) = x + l$, and for every $\Phi_k(1 < k \leqq n)$ we have either

1.

$$(x_2, \ldots, x_m)[\Phi_k(0, x_2, \ldots, x_m) = \Phi_p(O, x_2, \ldots, x_m)],$$
$$(x, x_2, \ldots, x_m)\{\Phi_\kappa[\Phi_i(x), x_2, \ldots, x_m] = \Phi_p, [x, \Phi_k(x, x_2, \ldots, x_m),$$

$$\tag{3.20}$$

$$x_2, \ldots, x_m]\}, \text{With } p, q < \kappa,\ ^{67}$$

Or

2.

$$(x_i, \ldots, x_m)[\Phi_k(x_i, \ldots, x_m) = \Phi_r(\Phi_i(\xi_i), \ldots, \Phi_{i_s}(\xi_s))],\ ^{68} \tag{3.21}$$
$$\text{with } r < k, i_v < k \text{ (for } r = 1, 2, \ldots, s),$$

or

3.

$$(x_1, \ldots, x_m)[\Phi_k(x_k, \ldots, x_m) = \Phi_1(\Phi_1, \ldots, \Phi_i(0))]. \tag{3.22}$$

We then form the propositions

$$(x)\overline{\Phi_1(x)} = 0\ \&\ (x, y)[\Phi_1(x) = \Phi_1(y) \to x = y], \tag{3.23}$$
$$(x)[\Phi_n(x) = 0]. \tag{3.24}$$

---

[65] Variables of the third kind may occur at all argument places occupied by individual variables, for example, $y = \varphi(x)$, $F(x, \varphi(y))$, $G(\psi(x, \varphi(y)), x)$, and the like.

[66] That is, by forming the conjunction.

In all of the formulas (18), (19), (20) (for $k = 2, 3, \ldots, n$) and in (21) and (22) we now replace the functions Φ_i by function variables φ_i and the number 0 by an individual variable x_0 not used so far, and we form the conjunction C of all the formulas thus obtained.

The formula $(Ex_0)C$ then has the required property that is,

1. If $(x)[\Phi(x) = 0]$ holds, $(Ex_Q)C$ is satisfiable. For the functions Φ_i, Φ_2, $\ldots$, Φ_r bviously yield a true proposition when substituted for φ_i, φ_2, $\ldots$, φ_n in $(Ex_0)C$);
2. If $(Ex_0)C$ is satisfiable. $(x)[\Phi(x) = 0]$ holds.

Proof. Let $\psi_i, \psi_2, \ldots, \psi_n$ be the functions (which exist by assumption) that yield a true proposition when substituted for $\varphi_i, \varphi_2, \ldots, \varphi_n$ in $(Ex_0)C$. Let $\mathfrak{I}$ be their domain of individuals. Since $(Ex_0)C$ holds for the functions ψ_i' there is an individual a (in $\mathfrak{I}$) such that all of the formulas (18)–(22) go over into true propositions, (18)–(22), when the Φ_i are replaced by the ψ_i and 0 by a. We now form the smallest snbclass of $\mathfrak{I}$ that contains a and is closed under the operation $\psi_i(x)$, This subclass ($\mathfrak{I}'$) has the property that every function ψ_1. when applied to elements of $\mathfrak{I}'$, again yields elements of $\mathfrak{I}'$. For this holds of ψ_i by the definition of $\mathfrak{I}'$, and by (18'), (19'), and (20') it carries over from ψ_i with smaller subscripts to ψ_i with larger ones. The functions that result from the ψ_i when these are restricted to the domain $\mathfrak{I}'$ of individuals will be denoted by ψ_i'. All of the formulas (18)–(22) hold for these functions also (when we replace 0 by a and Φ_i by ψ_i').

Because (21) holds for ψ_i' and a, we can map the individuals of $\mathfrak{I}'$ one-to-one onto the natural numbers in such a manner that a goes over into 0 and the function $\mathfrak{I}_1'$ into the successor function Φ_1. But by this mapping the functions ψ_i' go over into the functions Φ_1, and, since (22) holds for ψ_1' and a, $(x)[\Phi_n(x) = 0]$, that is, $(x)[\Phi(x) = 0]$, holds, which was to be proved.[69]

Since (for each particular F) the argument leading to Theorem X can be carried out in the system P, it follows that any proposition of the form $(x)F(x)$ (with recursive F) can in P be proved equivalent to the proposition that states about the corresponding formula of the r. f. c. that it is satisfiable. Hence the undecidability of one implies that of the other, which proves Theorem IX.[70]

---

[69] Theorem X implies, for example, that Fermat's problem and Goldbach's problem could be solved if the decision problem for the r. f. c. were solved.

[70] Theorem IX, of course, also holds for the axiom system of set theory and for its extensions by recursively definable ω-consistent classes of axioms, since there are undecidable propositions of the form $(x)F(x)$ (with recursive P) in these systems-too.

The results of Section 2 have a surprising consequence concerning a consistency proof for the system P (and its extensions), which can be stated as follows:

Theorem VII. *Let K be any recursive consistent[71] class of formulas; then the* SENTESTAIL FORMULA *stating that κ is consistent is not κ-PROVABLE; in particular, the consistency of P is not provable in P_i,[72] provided P is consistent (in the opposite case, of course, every proposition is provable [in P]).*

The proof (briefly outlined) is as follows. Let κ be some recursive class of FORMULAS chosen once and for all for the following discussion (in the simplest case it is the empty class). As appears from 1, page 608, only the consistency of k was used in proving that 17 Gen r is not κ-PROVABLE; of course, r (like p) depends on κ. that is, we have

$$\text{Wid}(k) \rightarrow \overline{Bew}_k(17 \text{ Gen } r), \tag{3.25}$$

that is, by (6.1),

$$\text{Wid}(k) \rightarrow (x)\overline{x \, B_k(17 \, \text{Gen} \, r)}.$$

By (13), we have

$$17 \text{ Gen } r = Sb(p^{19}_{z(p)}),$$

hence

$$\text{Wid}(\kappa) \rightarrow (x)\overline{x \, B_\kappa \, Sb(p^{19}_{z(p)})},$$

that is, by (8.1),

$$\text{Wid}(\kappa) \rightarrow (x)Q(x, p). \tag{3.26}$$

We now observe the following: all notions defined (or statements proved) in Section 2,[73] and in Section 4 up to this point, are also expressible (or provable) in P. For throughout we have used only the methods of definition and proof that are customary in classical mathematics, as they are formalized in the system P. In particular, κ (like every recursive class) is definable in P. Let w be the SENTESTIAL FORMULA by which Wid(κ) is expressed in P. According to (8.1), (9), and (10). the relation $Q(x, y)$ is expressed by the RELATION SIGN q, hence $Q(x, p)$ by r (since, by (12), $r = Sb(q^{19}_{z(p)})$), and the proposition $(x)Q(x, p)$ by 17 Gen r.

---

[71] "K is consistent" (abbreviatet by "Wid(K) is defined thus: Wid(k) $\equiv$ $(Ex)(\text{Form}(x)$ & $\overline{Bew}_K(x))$.

[72] This follows if we substitute the empty class of FORMULAS for K.

[73] From the definition of "recursive" on page 602 to the proof of Theorem VT inclusive.

Therefore, by (24), ω Imp (17 Gen r) is provable in P[74] (and a fortiori κ-PROVABLE). If now ω were κ-PROVABLE, then 17 Gen r would also be κ-PROVABLE, and from this it would follow, by (23), that κ is not consistent.

Let us observe that this proof, too, is constructive; that is, it allows us to actually derive a contradiction from κ, once a PROOF of ω from κ is given. The entire proof of Theorem XI carries over word for word to the axiom system of set theory, M, and to that of classical mathematics,[75] A, and here, too, it yields the result: There is no consistency proof for M, or for A, that could be formalized in M, or A, respectively, provided M, or A, is consistent. I wish to note expressly that Theorem XI (and the corresponding results for M and A) do not contradict Hilbert's formalistic viewpoint. For this viewpoint presupposes only the existence of a consistency proof in which nothing but finitary means of proof is used, and it is conceivable that there exist finitary proofs that *cannot* be expressed in the formalism of P (or of M or A).

Since, for any consistent class κ, ω is not κ-PROVABLE, there always are propositions (namely ω) that are undecidable (on the basis of κ) as soon as Neg(ω) is not κ-PROVABLE; in other words, we can, in Theorem VI, replace the assumption of ω-consistency by the following: The proposition "κ is inconsistent" is not κ-PROVABLE. (Note that there are consistent κ for which this proposition is κ-PROVABLE.)

In the present paper we have on the whole restricted ourselves to the system P, and we have only indicated the applications to other systems. The results will be stated and proved in full generality in a sequel to be published soon.[76] In that paper, also, the proof of Theorem XI, only sketched here, will be given in detail.

Note added 28 August 1953. In consequence of later advances, in particular of the fact that due to A. M. Turing's work[77] a precise and unquestionably adequate definition of the general notion of formal system[78] can now be

---

[74] That the truth of ω Imp (17 Gen r) can be inferred from (23) is simply due to the fact that the undecidable proposition 17 Gen r asserts its own unprovability, as was noted at the very beginning.

[75] See *von Neumann 1927*.

[76] [[This explains the "I" in the title of the paper. The author's intention was to publish this sequel in the next volume of the *Monaishefte*. The prompt acceptance of his results was one of the reasons that made him change his plan.]]

[77] See *Turing 1937*, p. 249.

[78] In my opinion the term "formal system" or "formalism" should never he used for anything but this notion. In a lecture at Princeton (mentioned *in Princeton University 1946*, p. 11 [[see *Davis 1965*, pp. 84—88]]) I suggested certain transfinite generalizations of formalisms, but these are something radically different from formal systems in the proper senso of the

given, a completely general version of Theorems VI and XI is now possible. That is, it can be proved rigorously that in *every* consistent formal system that contains a certain amount of finitary number theory there exist undecidable arithmetic propositions and that, moreover, the consistency of any such system cannot be proved in the system.

A.2 ON COMPLETENESS AND CONSISTENCY

(1931a)

Let Z be the formal system that we obtain by supplementing the Peano axioms with the schema of definition by recursion (on one variable) and the logical rules of the *restricted* functional calculus. Hence Z is to contain no variables other than variables for individuals (that is, natural numbers), and the principle of mathematical induction must therefore be formulated as a rule of inference. Then the following hold:

1. Given any formal system S in which there are finitely many axioms and in which the sole principles of inference are the rule of substitution and the rule of implication, if S contains[79] Z, S is incomplete, that is, there are

---

term, whose characteristic property is that reasoning in them, in principle, can be completely replaced by mechanical devices.

[79] That a formal system S contains another formal system T means that every proposition expressible (provable) in T is expressible (provable) also in S.

[[Remark by the author, 18 May 1966:]]

[This definition is not precise, and, if made precise in the straightforward manner, it does not yield a sufficient condition for the nondemonstrability in S of the consistency of S, A sufficient condition is obtained if one uses the following definition: "S contains T if and only if every meaningful formula (or axiom or rule (of inference, of definition, or of construction of axioms)) of T *is a* meaningful formula (or axiom, and so forth) of S, that is, if S is an extension of T".

Under the weaker hypothesis that Z is recursively one-to-one translatable into S, with demonstrability preserved in this direction, the consistency, even of very strong systems S, *may* be provable in S and even in primitive recursive number theory. However, what can be shown to be unprovable in S is the fact that the rules of the equational calculus applied to equations, between primitive recursive terms, demonstrable in S yield only correct numerical equations (provided that S possesses the property that is asserted to be unprovable). Note that it is necessary to prove this "outer" consistency of S (which for the usual systems is trivially equivalent with consistency) in order to "justify", in the sense of Hirbert's program, the transfinite axioms of a system S. ("Rules of the equational calculus" in the foregoing means the two rules of substituting primitive recursive terms for variables and substituting one such term for another to which it has been proved equal.)

The last-mentioned theorem and Theorem 1 of the paper remain valid for much weaker systems than Z. in particular for primitive recursive number theory, that is, what remains of

in$S$ propositions (in particular, propositions of Z) that are undecidable on the basis of the axioms of S, provided that S is ω-consistent. Here a system is said to be ω-consistent if, for no property F of natural numbers, $(Ex)\overline{Fx}$ as well as all the formulas $F(i)$, $i = 1, 2, \ldots$, are provable.

2. In particular, in every system S of the kind just mentioned the proposition that S is consistent (more precisely, the equivalent arithmetic proposition that we obtain by mapping the formulas one-to-one on natural numbers) is unprovable.

Theorems 1 and 2 hold also for systems in which there are infinitely many axioms and in which there are other principles of inference than those mentioned above, provided that when we enumerate the formulas (in order of increasing length and, for equal length, in lexicographical order) the class of numbers assigned to the axioms is definable and decidable [[entscheidungs-definit]] in the system Z, and that the same holds of the following relation $R(x_i, x_2, \ldots, x_n)$ between natural numbers: "the formula with number x_i follows from the formulas with numbers $x_2, \ldots, x_n$ by a single application of one of the rules of inference". Here a relation (class) $R(x_i, x_2, \ldots, x_n)$ is said to be decidable in Z if for every n-tuple $(k_i, k_2, \ldots, k_z)$ of natural numbers either $R(k_i, k_2, \ldots, k_n)$ or $\bar{R}(k_i, \ldots, k_n)$ is provable in Z. (At present no decidable number-theoretic relation is known that is not definable and decidable already in Z.)

If we imagine that the system Z is successively enlarged by the introduction of variables for classes of numbers, classes of classes of numbers, and so forth, together with the corresponding comprehension axioms, we obtain a sequence (continuable into the transfinite) of formal systems that satisfy the assumptions mentioned above, and it turns out that the consistency (ω-consistency) of any of those systems is provable in all subsequent systems. Also, the undecidable propositions constructed for the proof of Theorem I become decidable by the adjunction of higher types and the corresponding axioms; however, in the higher systems we can construct other undecidable propositions by the same procedure, and so forth. To be sure, all the propositions thus constructed are expressible in Z (hence are number-theoretic propositions); they are, however, not decidable in Z, but only in higher systems, for example, in that of analysis. In case we adopt a type-free construction of mathematics, as is done in the axiom system of set theory,

---

Z if quantifiers are omitted. With insignificant changes in the wording of the conclusions of the two theorems they even hold for any recursive translation into S of the equations between primitive recursive terms, under the sole hypothesis of ω-consistency (or outer consistency) of S in this translation.]

axioms of cardinality (that is axioms postulating the existence of sets of ever higher cardinality) take the place of the type extensions, and it follows that certain arithmetic propositions that are undecidable in Z become decidable by axioms of cardinality, for example, by the axiom that there exist sets whose cardinality is greater than every α_n, where $\alpha_0 = \aleph_0, \alpha_{z \div 1} = 2_n^\alpha$.

4

Appendix C

A Theory on Neurological Systems-Part I

By Dr. Bradley S. Tice

Abstract

In re-addressing points made by Penrose about Artificial Intelligence in his book *The Emperor's New Mind* (1989) a new view arises that gives a formal structure to a new model of neurological systems.

Introduction

In responding to statements by Roger Penrose in his book *The Emperor's New Mind* (1989) that deals with questions of Artificial Intelligence I hope to narrow some of these arguments and from that give substance to a new model of neurological systems. I will focus on three specific areas: 1. The Turing model of intelligence in computers, 2. Godel's Theorem and numbers and 3. Organic and non-organic systems.

Part I

In Alan Turing's paper "Computing Machinery and Intelligence" (1950) he considers the 'imitation game' as a test for whether a computer can pass for a human being as being a valid standard for a machine being intelligent. Penrose finds the Turing test of machine intelligence a 'valid indication of the presence of thought, intelligence, understanding, or consciousness is actually quite a strong one' (Penrose, 1989: 9) and 'Thus I am, as a whole, prepared to accept the Turing test as a roughly valid one in its chosen context' (Penrose, 1989: 10). Now while Penrose is clear to point out the weaknesses found

Bradley S. Tice, Language and Godels Theorem: A Revised Edition, 65–68.

in the Turing test of machine intelligence, such that a computer must be able to imitate a human but a human does not have to imitate a computer (Penrose, 1989: 9), he has still accepted it as a test of a machine's ability to have the following human properties: thought, intelligence, understanding, and consciousness.

In my paper "The Turing Machine: A Question of Linguistics?" I raise the issue that the Turing test of intelligence is just a question of language rather than intellect and even such factors as the type of language used, the physical and cultural abilities and knowledge of a human may cause a person to 'fail' this type of intelligence test (Tice, 1997/2004) Penrose has placed too many extraneous factors on this type of test and seems to miss the point that real intelligence is something beyond a game.

Part II

Penrose uses Godel's Theorem as a 'proof' that mathematical insight is by nature non-algorithmic (Penrose, 1989: 416). Unfortunately, Penrose has confused the fact that because Godel's Theorem states that not all axiomatic propositions can be proved, and hence, the thought process used for such thinking is non-algorithmic, the nature of mathematical 'insight', the action of realizing that a non-algorithmic process is itself a non-algorithmic process, gives light to the reason to suppose that human thought is non-algorithmic (Penrose, 1989: 417 and 429).

In my book *Formal Constraints to Formal Languages* (In Press) I address the question of Godel's Theorem and Hilbert's axiomatic foundations and that it did not provide an 'absolute' factor to the provability of propositions of number theory (Tice, In Press: 9). Also the use of a Universal Truth Machine [UTM] is given to present the basic procedure for Godel's Incompleteness Theorem (Tice, In Press: 13). An interesting result occurs when I substitute the words 'will never' with the word 'may' in the following sentence from the UTM:

UTM will never say G is true.

Resulting in the following sentence:

UTM may say G is true.

What results is that the Universal Truth Machine [UTM] becomes universal and changes the primary strengths found in Godel's Incompleteness Theorem, namely that some propositions can be axiomatically proved and some may not, but the robustness of the axiomatic system stays intact because it has accounted for such variants.

Part III

In von Neumann's paper "Probabilistic Logics and the Synthesis of Reliable Organisms from Unreliable Components" (1956/1963) he outlines the logical structure of reliable systems. In this paper he points out that the concept of 'error' must be viewed as not an extraneous and misdirected 'accident' but as an essential part of the process which is normal in the correct logical structure of such a system (Taub, 1963: 329). Von Neumann then proceeds to design automata that parallels a synthetic system, relays and circuits, with organic, neurons and the nervous system (Taub, 1963: 329).

An interesting point is if one considers numbers, all types, to be 'components' of a system and those that are functionally problematic, such as algorithms that 'maybe' provable but are, for some reason, intractable, become 'errors' within the system and hence, can be a unreliable part that can be replaced. In a system that is robust, numbers and algorithms that are 'problematic' are replaced by functional ones and allows the system to keep operating.

Summary

In summarizing this paper the following can be noted; that the Turing test for intelligence in machines is not viable, that Godel's Incompleteness Theorem is not an impediment to mechanical algorithmic functions if there is a selection of functional ones to replace the one's that are intractable. Organic systems, natural, may provide a systematic method for non-organic systems, mechanical, to follow in the function, rather than design, of a man-made system.

References

[1] Von Neumann, J. (1956/1963) "Probabilistic Logistics and the Synthesis of Reliable Organisms from Unreliable Components" in (ed. A.H. Taub) *John von Neumann Collected Works: Volume V*. New York: The Macmillan Company, pp. 329–378.

[2] Penrose, R. (1989) *The Emperor's New Mind: Concerning Computers, Minds, and The Laws of Physics*. Oxford, Oxford University Press.

[3] Tice, B.S. (1997) "The Turing Machine: A Question of Linguistics?" A paper given at the Pacific Division of the American Association for the Advancement of Science [AAAS] annual meeting at Oregon State University, June 22–26, 1997. A copy of the paper is reproduced in Appendix A of *Thought, Function, and Form: The Language of Physics*, B.S. Tice author, published by 1st Books Library, Indianapolis: 2004, pp. 207–214.

[4] Tice, B.S. (In press) *Formal Constraints to Formal Languages*. AuthorHouse, Indianapolis.

5

Appendix E

Minds, Machine and Godel

K. M. Sayre and F. J. Crosson

Minds, Machines and Gödel

First published in *Philosophy*, XXXVI, 1961, pp. (l12)–(127); reprinted in *The Modeling of Mind*, Kenneth M. Sayre and Frederick J. Crosson, eds., Notre Dame Press, 1963, pp. [269]–[270]; and *Minds and Machines*, ed. Alan Ross Anderson, Prentice-Hall, 1954, pp. {43}–{59}.

Gödel's theorem seems to me to prove that Mechanism is false, that is, that minds cannot be explained as machines. So also has it seemed to many other people: almost every mathematical logician I have put the matter to has confessed to similar thoughts, but has felt reluctant to commit himself definitely until he could see the whole argument set out, with all objections fully stated and properly met.[1] This I attempt to do.

Gödel's theorem states that in any consistent system which is strong enough to produce simple arithmetic there are formulae which cannot {44} be proved-in-the-system, but which we can see to be true. Essentially, we consider the formula which says, in effect, "This formula is unprovable-in-the-system". If this formula were provable-in-the-system, we should have a contradiction: for if it were provable-in-the-system, then it would not

---

[1] See A. M. Turing, "Computing Machinery and Intelligence," *Mind*, 1950, pp. 433–60, reprinted in *The World of Mathematics*, edited by James R. Newmann, pp. 2099–2123; and K. R. Popper, "Indeterminism in Quantum Physics and Classical Physics," *British Journal for Philosophy of Science*, 1 (1951), 179–88. The question is touched upon by Paul Rosenbloom; *Elements of Mathematical Logic*, pp. 207–8; Ernest Nagel and James R. Newmann, *Gödel's Proof*, pp. 100–2; and by Hartley Rogers, *Theory of Recursive Functions and Effective Computability* (mimeographed), 1957, Vol. 1, pp. 152 ff.

be unprovable-in-the-system, so that "This formula is unprovable-in-the-system" would be false: equally, if it were provable-in-the-system, then it would not be false, but would be true, since in any consistent system nothing false can be proved-in-the-system, but only truths. So the formula "This formula is unprovable-in-the-system" is not provable-in-the-system, but unprovable in-the-system. Further, if the formula "This formula is unprovable in-the-system" is unprovable-in-the-system, then it is true that that [256] formula is unprovable-in-the-system, that is, "This formula is unprovable-in-the-system" is true.

The foregoing argument is very fiddling, and difficult to grasp fully: it is helpful to put the argument the other way round, consider the possibility that "This formula is unprovable-in-the-system" might be false, show that that is impossible, and thus that the formula is true; whence it follows that it is unprovable. Even so, the argument remains persistently unconvincing: we feel that there must be a catch in it somewhere. The whole labour of Gödel's theorem is to show that there is no catch anywhere, and that the result can (113) be established by the most rigorous deduction; it holds for all formal systems which are (i) consistent, (ii) adequate for simple arithmetic—i.e., contain the natural numbers and the operations of addition and multiplication—and it shows that they are incomplete—i.e., contain unprovable, though perfectly meaningful, formulae, some of which, moreover, we, standing outside the system, can see to be true.

Gödel's theorem must apply to cybernetical machines, because it is of the essence of being a machine, that it should be a concrete instantiation of a formal system. It follows that given any machine which is consistent and capable of doing simple arithmetic, there is a formula which it is incapable of producing as being true—i.e., the formula is unprovable-in-the-system-but which we can see to be true. It follows that no machine can be a complete or adequate model of the mind, that minds are essentially different from machines.

We understand by a cybernetical machine an apparatus which performs a set of operations according to a definite set of rules. Normally we "programme" a machine: that is, we give it a set of instructions about what it is to do in each eventuality; and we feed in the initial "information" on which the machine is to perform its calculations. When we {45} consider the possibility that the mind might be a cybernetical mechanism we have such a model in view; we suppose that the brain is composed of complicated neural circuits, and that the information fed in by the senses is "processed" and acted upon or stored for future use. If it is such a mechanism, then given the way in which it is programmed—the way in which it is "wired up"—and the information which has been fed into it, the response—the "output"—is determined, and could, granted sufficient time, be calculated. Our idea of a machine is just this, that its behaviour is completely determined by the way it is made and the incoming "stimuli": there is no possibility of its acting on its own: given a certain form of construction and a certain input of information, then it must act in a certain specific way. We, however, shall be concerned not with what a machine *must* do, but with what it *can* do. That is, instead [257] of considering the whole set of rules which together determine exactly what a machine will do in given circumstances, we shall consider only an outline of those rules, which will delimit the possible responses of the machine, but not completely. The complete rules will determine the operations completely at every stage; at every stage there will be a definite instruction, e.g., "If the number is prime and greater than two add one and divide by two: if it is not prime, divide by its smallest factor": we, however, will consider the possibility of there being alternative instructions, e.g., "In a fraction you may divide top and bottom by *any* number which is a factor of both numerator and denominator". In thus (114) relaxing the specification of our model, so that it is no longer completely determinist, though still entirely mechanistic, we shall be able to take into account a feature often proposed for mechanical models of the mind, namely that they should contain a randomizing device. One could build a machine where the choice between a number of alternatives was settled by, say, the number of radium atoms to have disintegrated in a given container in the past half- minute. It is *prima facie* plausible that our brains should be liable to random effects: a cosmic ray might well be enough to trigger off a neural impulse. But clearly in a machine a randomizing device could not be introduced to choose any alternative whatsoever: it can only be permitted to choose between a number of allowable alternatives. It is all right to add any number chosen at random to both sides of an equation, but not to add one number to one side and another to the other. It is all right to choose to prove one theorem of Euclid rather than another, or to use one method rather than another, but not to "prove" something which is not true, or to use a "method of proof" which is not valid. Any {46} randomizing devices must allow choices only between those operations which will not lead to inconsistency: which is exactly what the relaxed specification of our model specifies Indeed, one might put it this way: instead of considering what a completely determined machine *must* do, we shall consider what a machine might be able to do if it had a randomizing device that acted whenever there were two or more operations possible, none of which could lead to inconsistency.

If such a machine were built to produce theorems about arithmetic (in many ways the simplest part of mathematics), it would have only a finite number of components, and so there would be only a finite number of types of operation it could do, and only a finite number of initial (115) assumptions it could operate on. Indeed, we can go further, and say that there would only be a *definite* number of types of operation, and of initial assumptions, that could be built into it. Machines are definite: anything which was indefinite or infinite we [258] should not count as a machine. Note that we say number of types of operation, not number of operations. Given sufficient time, and provided that it did not wear out, a machine could go on repeating an operation indefinitely: it is merely that there can be only a definite number of different sorts of operation it can perform.

If there are only a definite number of types of operation and initial assumptions built into the system, we can represent them all by suitable symbols written down on paper. We can parallel the operation by rules ("rules of inference" or "axiom schemata") allowing us to go from one or more formulae (or even from no formula at all) to another formula, and we can parallel the initial assumptions (if any) by a set of initial formulae ("primitive propositions", "postulates" or "axioms"). Once we have represented these on paper, we can represent every single operation: all we need do is to give formulae representing the situation before and after the operation, and note which rule is being invoked. We can thus represent on paper any possible sequence of operations the machine might perform. However long, the machine went on operating, we could, give enough time, paper and patience, write down an analogue of the machine's operations. This analogue would in fact be a formal proof: every operation of the machine is represented by the application of one of the rules: and the conditions which determine for the machine whether an operation can be performed in a certain situation, become, in our representation, conditions which settle whether a rule can be applied to a certain formula, i.e., formal conditions of applicability. Thus, construing our rules as rules of inference, we shall have a proof-sequence of {47} formulae, each one being written down in virtue of some formal rule of inference having been applied to some previous formula or formulae (except, of course, for the initial formulae, which are given because they represent initial assumptions built into the system). The conclusions it is possible for the machine to produce as being true will therefore correspond to the theorems that can be proved in the corresponding formal system. We now construct a Gödelian formula in this formal system. This formula cannot be*proved-in-the- system.* Therefore the machine cannot produce the corresponding formula as being true. But we can see that the Gödelian formula is true: any rational being could follow Gödel's argument, and convince himself that the Gödelian formula, although unprovable-in-the-system, was nonetheless—-in fact, for that very reason— true. Now any mechanical model of the mind must include a mechanism which can enunciate truths of arithmetic, because this is something which minds can do: in fact, it is easy to produce mechanical models which will in many respects produce truths of arithmetic far [259] better than human beings can. But in this one respect they cannot do so well: in that for every machine there is a truth which it cannot produce as being true, but which a mind can. This shows that a machine cannot be a complete and adequate model of the mind. It cannot do *everything* that a mind can do, since however much it can do, there is always something which it cannot do, and a mind can. This is not to say that we cannot build a machine to simulate any desired piece of mind-like behaviour: it is only that we cannot build a machine to simulate *every* piece of mind-like behaviour. We can (or shall be able to one day) build machines capable of reproducing bits of mind-like behaviour, and indeed of outdoing the performances of human minds: but however good the machine is, and however much better (116) it can do in nearly all respects than a human mind can, it always has this one weakness, this one thing which it cannot do, whereas a mind can. The Gödelian formula is the Achilles' heel of the cybernetical machine. And therefore we cannot hope ever to produce a machine that will be able to do all that a mind can do: we can never not even in principle, have a mechanical model of the mind.

This conclusion will be highly suspect to some people. They will object first that we cannot have it both that a machine *can* simulate *any* piece of mind-like behaviour, and that it *cannot* simulate *every* piece. To some it is a contradiction: to them it is enough to point out that there is no contradiction between the fact that for any natural number there can be produced a greater number, and the fact that a number cannot {48} be produced greater than every number. We can use the same analogy also against those who, finding a formula their first machine cannot produce as being true, concede that that machine is indeed inadequate, but thereupon seek to construct a second, more adequate, machine, in which the formula can be produced as being true. This they can indeed do: but then the second machine will have a Gödelian formula all of its own, constructed by applying Gödel's procedure to the formal system which represents its (the second machine's) own, enlarged, scheme of operations. And this formula the second machine will not be able to produce as being true, while a mind will be able to see that it is true. And if now a third machine is constructed, able to do what the second machine was unable to do, exactly the same will happen: there will be yet a third formula, the Gödelian formula for the formal system corresponding to the third machine's scheme of operations, which the third machine is unable to produce as being true, while a mind will still be able to see that it is true. And so it will go on. However complicated a machine we construct, it will, if it is a machine, correspond to a formal system, which in turn will be liable to the Gödel procedure [260] for finding a formula unprovable-in-that- system. This formula the machine will be unable to produce as being true, although a mind can see that it is true. And so the machine will still not be an adequate model of the mind. We are trying to produce a model of the mind which is mechanical—which is essentially "dead"—but the mind, being in fact "alive", can always go one better than any formal, ossified, dead, system can. Thanks to Gödel's theorem, the mind always has the last word.

A second objection will now be made. The procedure whereby the Gödelian formula is constructed is a standard procedure—only so could we be sure that a Gödelian formula can be constructed for every formal system. But if it is a standard procedure, then a machine should be able to be programmed to carry it out too. We could construct a machine with the usual operations, and in addition an (117) operation of going through the Gödel procedure, and then producing the conclusion of that procedure as being true; and then repeating the procedure, and so on, as often as required. This would correspond to having a system with an additional rule of inference which allowed one to add, as a theorem, the Gödelian formula of the rest of the formal system, and then the Gödelian formula of this new, strengthened formal system, and so on. It would be tantamount to adding, to the original formal system an infinite sequence of axioms, each the Gödelian formula of the system hitherto obtained. Yet even so, the matter is not settled: for the machine with a Gödelizing {49} operator, as we might call it, is a *different* machine from the machines without such an operator; and, although the machine with the operator would be able to do those things in which the machines without the operator were outclassed by a mind, yet we might expect a mind, faced with a machine that possessed a Gödelizing operator, to take this into account, and out-Gödel the new machine, Gödelizing operator and all. This has, in fact, proved to be the case. Even if we adjoin to a formal system the infinite set of axioms consisting of the successive Gödelian formulae, the resulting system is still incomplete, and contains a formula which cannot be proved-in-the-system, although a rational being can, standing outside the system, see that it is true.[2] We had expected this, for even if an infinite set of axioms were added, they would have to be specified by some finite rule or specification, and this further rule or specification could then be taken into account by a mind considering the enlarged formal system. In a sense, just because the mind has the last word, it can always pick a hole in any formal system presented to it as a model of its own workings. The [261] mechanical model must be, in some sense, finite and definite: and then the mind can always go one better.

This is the answer to one objection put forward by Turing.[3] He argues that the limitation to the powers of a machine do not amount to anything much. Although each individual machine is incapable of getting the right answer to some questions, after all each individual human being is fallible also: and in any case "our superiority can only be felt on such an occasion in relation to the one machine over which we have scored our petty triumph. There would be no question of triumphing simultaneously over *all* machines." But this is not the point. We are not discussing whether machines or minds are superior, but whether they are the same. In some respect machines are undoubtedly superior to human minds; and the question on which they are stumped is admittedly, a rather niggling, even (118) trivial, question. But it is enough, enough to show that the machine is *not the same* as a mind. True, the machine can do many things that a human mind cannot do: but if there is of necessity something that the machine cannot do, though the mind can, then, however trivial the matter is, we cannot equate the two, and cannot hope ever to have a mechanical model that will adequately represent the mind. Nor does it signify that it is only an individual machine we have triumphed over: for the triumph is not over only *an* individual machine, but over *any* individual that anybody cares to specify—in Latin {50} *quivis* or *quilibet*, not *quidarm*—and   a mechanical model of a mind must be an individual machine. Although it is true that any particular "triumph" of a mind over a machine could be "trumped" by another machine able to produce the answer the first machine could not produce, so that "there is no question of triumphing simultaneously over all machines", yet this is irrelevant. What is at issue is not the unequal contest between one mind and all machines, but whether there could be any, single, machine that could do all a mind can do. For the mechanist thesis to hold water, it must be possible, in principle, to produce a model, a single model, which can do everything the mind can do. It is like a

---

[2] Gödel's original proof applies; $v.^{\perp}$ I *init.* and $^{\perp}$ 6 *init.* of his Lectures at the Institute of Advanced Study, Princeton, N.J., U.S.A., 1934.

[3] *Mind*, 1950, pp. 444–5; Newman, p. 2110.

game.[4] The mechanist has first turn. He produces *a—any*, but only a *definite one*—mechanical model of the mind. I point to something that it cannot do, but the mind can. The mechanist is free to modify his example, but each time he does so, I am entitled to look for defects in the revised model. If the mechanist can devise a model that I cannot find fault with, his [262] thesis is established: if he cannot, then it is not proven: and since—as it turns out—he necessarily cannot, it is refuted. To succeed, he must be able to produce some definite mechanical model of the mind—anyone he likes, but one he can specify, and will stick to. But since he cannot, in principle cannot, produce any mechanical model that is adequate, even though the point of failure is a minor one, he is bound to fail, and mechanism must be false.

Deeper objections can still be made. Gödel's theorem applies to deductive systems, and human beings are not confined to making only deductive inferences. Gödel's theorem applies only to consistent systems, and one may have doubts about how far it is permissible to assume that human beings are consistent. Gödel's theorem applies only to formal systems, and there is no *a priori* bound to human ingenuity which rules out the possibility of our contriving some replica of humanity which was not representable by a formal system.

Human beings are not confined to making deductive inferences, and it has been urged by C.G. Hempel[5] and Hartley Rogers[6] that a fair model of the mind would have to allow for the possibility of making non-deductive inferences, and these might provide a way of escaping the Gödel result. Hartley Rogers makes the specific suggestion that the {51} machine should be programmed to entertain various propositions which had not been proved or disproved, and on occasion to add them to its list of axioms. Fermat's last theorem or Goldbach's conjecture might thus be added. If subsequently their inclusion was found to lead to a contradiction, they would be dropped again, and indeed in those circumstances their negations would be added to the list of theorems. In this sort of way a machine might well be constructed which was able to produce as true certain formulae which could not be proved from its axioms according to its rules of inference. And therefore the method of demonstrating the mind's superiority over the machine might no longer work.

---

[4] For a similar type of argument, see J. R. Lucas: "The Lesbian Rule"; *Philosophy*, (July 1955) pp. 202–206; and "On Not Worshipping Facts"; *The Philosophical Quarterly*, April 1958, p. 144.

[5] In private conversation.

[6] *Theory of Recursive Functions and Effective Computability*, 1957, Vol. 1, pp. 152 ff.

The construction of such a machine, however, presents difficulties. It cannot accept all unprovable formulae, and add them to its axioms, or it will find itself accepting both the Gödelian formula and its negation, and so be inconsistent. Nor would it do if it accepted the first of each pair of undecidable formulae, and, having added that to its axioms, would no longer regard its negation as undecidable, and so would never accept it too: for it might happen on the wrong member of the pair: it might accept the negation of the Gödelian formula rather than the Gödelian formula itself. And the system constituted [263] by a normal set of axioms with the negation of the Gödelian formula adjoined, although not inconsistent, is an unsound system, not admitting of the natural interpretation. It is something like non-Desarguian geometries in two dimensions: not actually inconsistent, but rather wrong, sufficiently much so to disqualify it from serious consideration. A machine which was liable to infelicities of that kind would be no model for the human mind.

It becomes clear that rather careful criteria of selection of unprovable formulae will be needed. Hartley Rogers suggests some possible ones. But once we have rules generating new axioms, even if the axioms generated are only provisionally accepted, and are liable to be dropped again if they are found to lead to inconsistency, then we can set about doing a Gödel on this system, as on any other. We are in the same case as when we had a rule generating the infinite set of Gödelian formulae as axioms. In short, however a machine is designed, it must proceed either at random or according to definite rules. In so far as its procedure is random, we cannot outsmart it: (120) but its performance is not going to be a convincing parody of intelligent behaviour: in so far as its procedure is in accordance with definite rules, the Gödel method can {52} be used to produce a formula which the machine, according to those rules, cannot assert as true, although we, standing outside the system, can see it to be true.[7]

Gödel's theorem applies only to consistent systems. All that we can prove *formally* is that *if* the system is consistent, then the Gödelian formula is unprovable-in-the-system. To be able to say categorically that the Gödelian formula is unprovable-in- the-system, and therefore true, we must not only be dealing with a consistent system, but be able to say that it is consistent. And, as Gödel showed in his second theorem—a corollary of his first—it

---

[7] Gödel's original proof applies if the rule is such as to generate a primitive recursive class of additional formulae; $v.^{\perp}$ I *init.* and $^{\perp}$ 6 *init.* of his Lectures at the Institute of Advanced Study, Princeton, N.J., U.S.A., 1934. It is in fact sufficient that the class be recursively enumerable. See Barkley Rosser: "Extensions of some theorems of Gödel and Church," *Journal of Symbolic Logic*, **1**, 1936, pp. 87–91.

is impossible to prove in a consistent system that that system is consistent. Thus in order to fault the machine by producing a formula of which we can say both that it is true and that the machine cannot produce it as true, we have to be able to say that the machine (or, rather, its corresponding formal system) is consistent; and there is no absolute proof of this. All we can do is to examine the machine and see if it appears consistent. There always remains the possibility of some inconsistency not yet detected. At best we can say that the machine is consistent, provided we are. But by what right can we do this? Gödel's second [264] theorem seems to show that a man cannot assert his own consistency, and so Hartley Rogers[8] argues that we cannot really use Gödel's first theorem to counter the mechanist thesis unless we can say that "there are distinctive attributes which enable a human being to transcend this last limitation and assert his own consistency while still remaining consistent".

A man's untutored reaction if his consistency is questioned is to affirm it vehemently: but this, in view of Gödel's second theorem, is taken by some philosophers as evidence of his actual inconsistency. Professor Putnam[9] has suggested that human beings are machines, but inconsistent machines. If a machine were wired to correspond to an inconsistent system, then there would be no well-formed formula which it could not produce as true; and so in no way could it be proved to be inferior to a human being. Nor could we make its inconsistency a reproach to it—are not men inconsistent too? Certainly women are, and politicians; and {53} even male non-politicians (121) contradict themselves sometimes, and a single inconsistency is enough to make a system inconsistent.

The fact that we are all sometimes inconsistent cannot be gainsaid, but from this it does not follow that we are tantamount to inconsistent systems. Our inconsistencies are mistakes rather than set policies. They correspond to the occasional malfunctioning of a machine, not its normal scheme of operations. Witness to this that we eschew inconsistencies when we recognize them for what they are. If we really were inconsistent machines, we should remain content with our inconsistencies, and would happily affirm both halves of a contradiction. Moreover, we would be prepared to say absolutely anything— which we are not. It is easily shown[10] that in an inconsistent formal system everything is provable, and the requirement of consistency turns out to be

---

[8] Op. cit., p. 154.

[9] University of Prineeton, N.J., U.S.A. in private conversation.

[10] See, e.g., Alonzo Church: *Introduction to Mathematical Logic*, Princeton, Vol.1, $\perp$ 17, p. 108.

just that not everything can be proved in it—it is not the case that "anything goes." This surely is a characteristic of the mental operations of human beings: they are selective: they do discriminate between favoured—true—and unfavoured—false—statements: when a person is prepared to say anything, and is prepared to contradict himself without any qualm or repugnance, then he is adjudged to have "lost his mind". Human beings, although not perfectly consistent, are not so much inconsistent as fallible.

A fallible but self-correcting machine would still be subject to Gödel's results. Only a fundamentally inconsistent machine would [265] escape. Could we have a fundamentally inconsistent, but at the same time self- correcting machine, which both would be free of Gödel's results and yet would not be trivial and entirely unlike a human being? A machine with a rather *recherché*: inconsistency wired into it, so that for all normal purposes it was consistent, but when presented with the Gödelian sentence was able to prove it?

There are all sorts of ways in which undesirable proofs might be obviated. We might have a rule that whenever we have proved p and not-p, we examine their proofs and reject the longer. Or we might arrange the axioms and rules of inference in a certain order, and when a proof leading to an inconsistency is proffered, see what axioms and rules are required for it, and reject that axiom or rule which comes last in the ordering. In some such way as this we could have an inconsistent system, with a stop-rule, so that the inconsistency was never allowed to come out in the form of an inconsistent formula.

The suggestion at first sight seems attractive: yet there is something deeply wrong. Even though we might preserve the facade of consistency {54} by having a rule that whenever two inconsistent formulae (122) appear we were to reject the one with the longer proof, yet such a rule would be repugnant in our logical sense. Even the less arbitrary suggestions are too arbitrary. No longer does the system operate with certain definite rules of inference on certain definite formulae. Instead, the rules apply, the axioms are true, provided ... we do not happen to find it inconvenient. We no longer know where we stand. One application of the rule of Modus Ponens may be accepted while another is rejected: on one occasion an axiom may be true, or another apparently false. The system will have ceased to be a formal logical system, and the machine will barely qualify for the title of a model for the mind. For it will be far from resembling the mind in its operations: the mind does indeed try out dubious axioms and rules of inference; but if they are found to lead to contradiction, they are rejected altogether. We try out axioms and rules of inference provisionally—true: but we do not keep them, once they are found to lead to contradictions. We may seek to replace them with others, we may feel that our formalization is at fault, and that though some axiom or rule of inference of this sort is required, we have not been able to formulate it quite correctly: but we do not retain the, faulty formulations without modification, merely with the proviso that when the argument leads to a contradiction we refuse to follow it. To do this would be utterly irrational. We should be in the position that on some occasions when supplied with the premises of a Modus Ponens, say, we applied the rule and allowed the conclusion, and [266] on other occasions we refused to apply the rule, and disallowed the conclusion. A person, or a machine, which did this without being able to give a good reason for so doing, would be accounted arbitrary and irrational. It is part of the concept of "arguments" or "reasons" that they are in some sense general and universal: that if Modus Ponens is a valid method of arguing when I am establishing a desired conclusion, it is a valid method also when you, my opponent, are establishing a conclusion I do not want to accept. We cannot pick and choose the times when a form of argument is to be valid; not if we are to be reasonable. It is of course true, that with our informal arguments, which are not fully formalized, we do distinguish between arguments which are at first sight similar, adding further reasons why they are nonetheless not really similar: and it might be maintained that a {55} machine might likewise be entitled to distinguish between arguments at first sight similar, if it had good reason for doing so. And it might further be maintained that the machine had good reason for rejecting those patterns of argument it did reject, indeed the best of reasons, namely the avoidance of contradiction. But that, if it is a reason at all, is too good a reason. We do not lay it to a man's credit that he avoids contradiction merely by refusing to accept those arguments which would lead him to it, for no other (123) reason than that otherwise he would be led to it. Special pleading rather than sound argument is the name for that type of reasoning. No credit accrues to a man who, clever enough to see a few moves of argument ahead, avoids being brought to acknowledge his own inconsistency, by stonewalling as soon as he sees where the argument will end. Rather, we account him inconsistent too, not, in his case, because he affirmed and denied the same proposition, but because he used and refused to use the same rule of inference. A stop-rule on actually enunciating an inconsistency is not enough to save an inconsistent machine from being called inconsistent.

The possibility yet remains that we are inconsistent, and there is no stop-rule, but the inconsistency is so *recherché:* that it has never turned up. After all, naive set-theory, which was deeply embedded in common- sense ways of thinking did turn out to be inconsistent. Can we be sure that a similar fate is not in store for simple arithmetic too? In a sense we cannot, in spite of our great feeling of certitude that our system of whole numbers which can be added and multiplied together is never going to prove inconsistent. It is just conceivable we might find we had formalized it incorrectly. If we had, we should try and formulate anew our intuitive concept of number, as we have our intuitive concept of a set. If we did this, we should of course recast our system: our present axioms and rules of inference would [267] be utterly rejected: there would be no question of our using and not using them in an "inconsistent" fashion. We should, once we had recast the system, be in the same position as we are now, possessed of a system believed to be consistent, but not provably so. But then could there not be some other inconsistency? It is indeed a possibility. But again no inconsistency once detected will be tolerated. We are determined not to be inconsistent, and are resolved to root out inconsistency, should any appear. Thus, although we can never be completely certain or completely free of the risk of having to think out our mathematics again, the ultimate position must be one of two: either we have a system of simple arithmetic which to the best of our knowledge and belief is consistent: or there is no such system possible. In the former case we are in the same position as at present: in the {56} latter, if we find that no system containing simple arithmetic can be free of contradictions, we shall have to abandon not merely the whole of mathematics and the mathematical sciences, but the whole of thought.

It may still be maintained that although a man must in this sense assume, he cannot properly affirm, his own consistency without thereby belying his words. We may be consistent; indeed we have every reason to hope that we are: but a necessary modesty forbids us from saying so. Yet this is not quite what Gödel's second theorem states. Gödel has shown that in a consistent system a formula (124) stating the consistency of the system cannot be proved *in that system*. It follows that a machine, if consistent, cannot produce as true an assertion of its own consistency: hence also that a mind, *if it were really a machine*, could not reach the conclusion that it was a consistent one. For a mind which is not a machine no such conclusion follows. All that Gödel has proved is that a mind cannot produce a formal proof of the consistency of a formal system inside the system itself: but there is no objection to going outside the system and no objection to producing informal arguments for the consistency either of a formal system or of something less formal and less systematized. Such informal arguments will not be able to be completely formalized: but then the whole tenor of Gödel's results is that we ought not to ask, and cannot obtain, complete formalization. And although it would have been nice if we could have obtained them, since completely formalized arguments are more coercive than informal ones, yet since we cannot have all our arguments cast into that form, we must not hold it against informal arguments that they are informal or regard them all as utterly worthless. It therefore seems to me both proper and reasonable for a mind to assert its own consistency: proper, because although machines, as we might have expected, are [268] unable to reflect fully on their own performance and powers, yet to be able to be self-conscious in this way is just what we expect of minds: and reasonable, for the reasons given. Not only can we fairly say simply that we *know* we are consistent, apart from our mistakes, but we must in any case *assume* that we are, if thought is to be possible at all; moreover we are selective, we will not, as inconsistent machines would, say anything and everything whatsoever: and finally we can, in a sense, *decide* to be consistent, in the sense that we can resolve not to tolerate inconsistencies in our thinking and speaking, and to eliminate them, if ever they should appear, by withdrawing and cancelling one limb of the contradiction.

We can see how we might almost have expected Gödel's theorem to distinguish self-conscious beings from inanimate objects. The essence of {57} the Gödelian formula is that it is self-referring. It says that "This formula is unprovable-in-this-system". When carried over to a machine, the formula is specified in terms which depend on the particular machine in question. The machine is being asked a question about its own processes. We are asking it to be self-conscious, and say what things it can and cannot do. Such questions notoriously lead to paradox. At one's first and simplest attempts to philosophize, one becomes entangled in questions of whether when one knows something one knows that one knows it, and what, when one is thinking of oneself, is being thought about, and what is doing the thinking. After one has been puzzled and bruised by this (125) problem for a long time, one learns not to press these questions: the concept of a conscious being is, implicitly, realized to be different from that of an unconscious object. In saying that a conscious being knows something, we are saying not only that he knows it, but that he knows that he knows it, and that he knows that he knows that he knows it, and so on, as long as we care to pose the question: there is, we recognize, an infinity here, but it is not an infinite regress in the bad sense, for it is the questions that peter out, as being pointless, rather than the answers. The questions are felt to be pointless because the concept contains within itself the idea of being able to go on answering such questions indefinitely. Although conscious beings have the power of going on, we do not wish to exhibit this simply as a succession of tasks they are able to perform, nor do we see the mind as an infinite sequence of selves and super-selves and super-superselves. Rather, we insist that a conscious being is a unity, and though we talk about parts of the mind, we do so only as a metaphor, and will not allow it to be taken literally.

The paradoxes of consciousness arise because a conscious being can be aware of itself, as well as of other things, and yet cannot [269] really be construed as being divisible into parts. It means that a conscious being can deal with Gödelian questions in a way in which a machine cannot, because a conscious being can both consider itself and its performance and yet not be other than that which did the performance. A machine can be made in a manner of speaking to "consider" its own performance, but it cannot take this "into account" without thereby becoming a different machine, namely the old machine with a "new part" added. But it is inherent in our idea of a conscious mind that it can reflect upon itself and criticize its own performances, and no extra part is required to do this: it is already complete, and has no Achilles' heel.

The thesis thus begins to become more a matter of conceptual analysis {58} than mathematical discovery. This is borne out by considering another argument put forward by Turing.[11] So far, we have constructed only fairly simple and predictable artefacts. When we increase the complexity of our machines there may, perhaps, be surprises in store for us. He draws a parallel with a fission pile. Below a certain "critical" size, nothing much happens: but above the critical size, the sparks begin to fly. So too, perhaps, with brains and machines. Most brains and all machines are, at present, "subcritical"—they react to incoming stimuli in a stodgy and uninteresting way, have no ideas of their own, can produce only stock responses—but a few brains at present, and possibly some machines in the future, are super-critical, and scintillate on their own account. (126) Turing is suggesting that it is only a matter of complexity, and that above a certain level of complexity a qualitative difference appears, so that 44 super-critical" machines will be quite unlike the simple ones hitherto envisaged.

This may be so. Complexity often does introduce qualitative differences. Although it sounds implausible, it might turn out that above a certain level of complexity, a machine ceased to be predictable, even in principle, and started doing things on its own account, or, to use a very revealing phrase, it might begin to have a mind of its own. It might begin to have a mind of its own.

---

[11] *Mind*, 1950, p. 454; Newman, pp. 2117–18.

It would begin to have a mind of its own when it was no longer entirely predictable and entirely docile, but was capable of doing things which we recognized as intelligent, and not just mistakes or random shots, but which we had not programmed into it. But then it would cease to be a machine, within the meaning of the act. What is at stake in the mechanist debate is not how minds are, or might be, brought into being, but how they operate. It is essential for the mechanist thesis that the mechanical model of the mind shall operate according [270] to "mechanical principles", that is, that we can understand the operation of the whole in terms of the operations of its parts, and the operation of each part either shall be determined by its initial state and the construction of the machine, or shall be a random choice between a determinate number of determinate operations. If the mechanist produces a machine which is so complicated that this ceases to hold good of it, then it is no longer a machine for the purposes of our discussion, no matter how it was constructed. We should say, rather, that he had created a mind, in the same sort of sense as we procreate people at present. There would then be two ways of bringing new minds into the world, the traditional way, by begetting children born of women, and a new way by constructing very, very complicated systems of, say, valves {59} and relays. When talking of the second way, we should take care to stress that although what was created looked like a machine, it was not one really, because it was not just the total of its parts. One could not tell what it was going to do merely by knowing the way in which it was built up and the initial state of its parts: one could not even tell the limits of what it could do, for even when presented with a Gödel-type question, it got the answer right. In fact we should say briefly that any system which was not floored by the Gödel question was *eo ipso* not a Turing machine, i.e., not a machine within the meaning of the act.

If the proof of the falsity of mechanism is valid, it is of the greatest consequence for the whole of philosophy. Since the time of Newton, the bogey of mechanist determinism has obsessed philosophers. If we were to be scientific, it seemed that we must look on human beings as (127) determined automata, and not as autonomous moral agents; if we were to be moral, it seemed that we must deny science its due, set an arbitrary limit to its progress in understanding human neurophysiology, and take refuge in obscurantist mysticism. Not even Kant could resolve the tension between the two standpoints. But now, though many arguments against human freedom still remain, the argument from mechanism, perhaps the most compelling argument of them all, has lost its power. No longer on this count will it be incumbent on the natural philosopher to deny freedom in the name of science:

no longer will the moralist feel the urge to abolish knowledge to make room for faith. We can even begin to see how there could be room for morality, without its being necessary to abolish or even to circumscribe the province of science. Our argument has set no limits to scientific enquiry: it will still be possible to investigate the working of the brain. It will still be possible to produce mechanical models of the mind. Only, now we can see that no mechanical model will be completely adequate, nor any explanations [271] in purely mechanist terms. We can produce models and explanations, and they will be illuminating: but, however far they go, there will always remain more to be said. There is no arbitrary bound to scientific enquiry: but no scientific enquiry can ever exhaust the infinite variety of the human mind.[12]

See also *Turn Over the Page* a talk I gave on 25/5/96 at a BSPS conference in Oxford and *"Minds, Machines and Gödel: A Retrospect"*, in P.J.R. Millican and A.Clark, eds., *Machines and Thought: The Legacy of Alan Turing*, Oxford, 1996, pp. 103–124.

A full discussion of the issues raised is now available in*Etica e Politica, 2003.*

GoogleScholar gives a large number (ca. 242) of references to critical discussions. an old list of criticisms and discussions of the Gödelian argument. Click here to return to home page Click here to return to bibliography

---

[12] Some objections by Benacerraf and Putman are considered in *"Satan Stuhitled"* A fuller account, in which further objections are considered, is given in J.R.Lucas, *The Freedom of the Will*, Oxford, 1970. Most critics concentrate their fire on "Minds, Machines and Gödel", without looking at *The Freedom of the Will*. In recent years it has been out of print. But under a new intitative by OUP, it is now available again. Single copies are printed on a one-off basis. I commend it to those who think there are holes in the article reprinted here. (In the fulness of time I hope to scan the relevant pages and make them available on the Web: but I have many other pressing calls on my time.)

Index

Lightning Source UK Ltd.
Milton Keynes UK
UKOW07f1034171214

243278UK00001B/2/P